Past
Forgetting

Past Forgetting

My Memory Lost and Found

Jill Robinson

Cliff Street Books

An Imprint of HarperCollins*Publishers*

A hardcover edition of this book was published by Cliff Street Books, an imprint of HarperCollins Publishers, in 1999.

HarperCollins books may be purchased for educational, business, or sales promotional use. For information please write: Special Markets Department, HarperCollins Publishers Inc., 10 East 53rd Street, New York, NY 10022.

First paperback edition published 2000.

Designed by Liane F. Fuji

The Library of Congress has catalogued the hardcover edition as follows:
    Robinson, Jill.
        Past forgetting: my memory lost and found/Jill
    Robinson—1st ed.
            p. cm.
        ISBN 0-06-019430-8
        1. Robinson, Jill—Health 2. Women novelists, American—20th century Biography. 3. Epilepsy—Patients—England—London Biography. 4. Amnesia—Patients—England—London Biography. 5. Memory. I. Title.
    PS3568.028915Z472 1999
    813'.54—dc21   99-34975

ISBN 0-06-093234-1

00 01 02 03 04 ❖/HC 10 9 8 7 6 5 4 3 2 1

FOR
STUART SHAW

I'll see you again.
Whenever Spring breaks through again;
Time may lie heavy between,
But what has been
Is past forgetting

—NOEL COWARD, *I'LL SEE YOU AGAIN*

Live as if you were living already for the second time. . . . There is
nothing which will stimulate a man's sense of responsibleness more
than this maxim, which invites him to imagine first, that the present
is past, and second that the past may yet be changed and amended.
Such a precept confronts him with life's finiteness, as well as the
finality of what he makes out of both his life and himself.

—VIKTOR F. FRANKL, *MAN'S SEARCH FOR MEANING*

Writers look for order and pattern in memory, fitting things together,
giving inner coherence to our lives.

—DORIS LESSING

Acknowledgments

I wish to thank the following authors and publishers for granting their kind permission:

Alan Baddeley, *Your Memory, a User's Guide* (Penguin Books, 1983)

Victor E. Frankl, *Man's Search for Meaning* (Simon & Schuster, 1959)

Barry Gordon, *Memory, Remembering and Forgetting in Everyday Life* (Mastermedia Limited, 1995)

Donlyn Lyndon and Charles W. Moore, *Chambers for a Memory Palace* (MIT Press, 1994)

Steven Rose, *The Making of Memory* (Bantam Press, 1992)

Daniel Schacter, *Searching for Memory* (Basic Books, 1996)

Dore Schary, *Heyday, An Autobiography* (Little, Brown and Company, 1979)

M.R. Trimble, *Women and Epilepsy* (1991)

D. Upton, P. Thompson and R. Corcorau, *Cognitive Differences between Males and Females with Epilepsy*

Peter Wolson, *The Vital Role of Adaptive Grandiosity in Artistic Creativity* (Psychoanalytic Review, 82[4], August 1995)

Frances A. Yates, *The Art of Memory* (ARK Paperbacks, 1984; orig. 1966)
    And, Dr. K. J. Zilkha MD, FRCP.
And with gratitude for their continuing interest and encouragement to
                    Ted and Vada Stanley
                            &
                    Bruce and Lueza Gelb.
    And also to Leon and Karen Allen, Martha Greene, Rob and Sherry
Johnson, Rosalie Mishkin, and Steve Burn.
    And, of course, to
                        Lynn Nesbit
                            &
                    Diane Reverand.
    And my thanks to Mary Carwile.
    And to Matthew Guma.
    My affection to Laurie Lipton, Corinne Laurie, Charles and Angie
Montagu, Susan Granger, and Joanne and Gil Segel.
    And always to Joy Schary and Jeb Schary.
And my love to Stuart Shaw, Jr., Susan Shaw, and Philip Shaw.
    Above all, my love and appreciation to Jeremy Zimmer and Johanna
Simmel for the example of initiative, resolution, and spirit they've shown
in the creation of their lives.

C · H · A · P · T · E · R

1

*I*t begins like this. I am awake. Sunlight comes through the window. A warm body sits next to me on the bed. A firm torso or arm, pressing close, male or female. Not sure. The sun frames the blond hair. Solid presence–stability.

Crisp, heavy sheets. This is a hospital somewhere. And I'm in it.

Looking hard against the sun, I can see the face. "So, you're captivating," I say, "and who are you?"

"I'm your husband," he says. "I'm Stuart."

"That's a beginning," I laugh, "and who am I?" I'd like to ask, "And where are we," but that's too much to know just now.

"You're Jill. You're a writer."

He's scared, I can hear that. But at the same time he's able to be reassuring. "You're going to be just fine."

I can tell he's lying. "I didn't ask."

He takes my hand. "But you will. Do you remember climbing out of the pool?"

"Yes," I say. Can he tell when I'm lying?

I remember swimming. I am stretching every muscle to match the drive of the woman leading aerobics before this—collar blade outlined in sweat, thigh bones shadowed like the ridge under Ava Gardner's brows. Am I in Palm Springs? Twenty laps. I've got nineteen. "You're there," I urge myself on, "just one more."

Now I'm here in this hospital. "So, did I hit my head?" I don't wait for an answer. "How long have I been out?"

"A while," he says.

I lean against his arm.

It is night. Someone brawny is sitting beside me. "Hello . . ."—male voice—"now have some soup." He tries to feed me. I can't taste the name of it. "You could have drowned," he's telling me, "but you got out of the pool somehow."

I touch his forearm lightly. "This is very patient, nice of you to sit here with me."

"I'm your husband."

"I know—but I don't know." Tears. "If you know what I mean."

Here's the next banner of time I catch hold of. A man in white comes in. He puts up the shade. Behind him the world is flat, plain, and soft green, like a land in an old-fashioned children's book. I've never been out of America. Our family doesn't fly. Not since Carole Lombard was killed in the plane crash.

"It's sweet outside," I say.

"Are you asking where you are?"

"Probably."

He's great looking. Blond, built like Spencer Tracy, he'll charm my father. "Are you my doctor?" I look him up and down.

He sits down next to me on the bed and puts his hand over mine. "I'm your husband, Jill. You're at the Stoke Mandeville Hospital near Tring."

Mandeville. *That's familiar,* I think.

"In England," he says.

"You know, I thought England looked like this." I don't want to ask how I got here. Not yet. My head's splitting. "The kids—is there a phone

I can use? I have to call my kids." I look him over. I can't make any connection here between him and my kids, Jeremy and Johanna. "Who's with them?"

"Jill." He's trying to get me to pay attention, to see how long I can hold onto what he tells me. "You've been in a coma and you've forgotten things. It's only temporary." He hopes. "Jeremy and Johanna are grown."

"I've been out so long?" I try to sit up.

"It will all come back." He's trying to reassure both of us. He sounds English. My life is over and I've come back in an old war movie.

"There's a piece missing," I explain. I touch his hand and I'm gone again.

I'm leaning against a man's chest. I'm seriously unhappy. "You'll remember soon."

Anger is the last thing I remembered before the seizure, which put me into the coma, shattered my ability to remember and erased years of recollection.

My husband and I had come to a spa outside London, where we live. We made a pact to get fit, he reminds me now.

"You didn't want to rush dressing to go in for breakfast. You were trying on outfits. So I went back to bed to read until you decided you were ready. Then you discovered I'd brought chocolate with me to the spa. You put on your bathing suit, said, 'Forget breakfast, I'm going swimming.'"

"So I was the one who was angry."

"Yes."

Is his voice appealing because it's familiar, or because the tone, the rhythm, appeals to my taste? Do I remember my taste? What if I've forgotten how I dress? It's okay. He won't mind. Maybe.

We're searching each other's faces. He's trying to see if I don't remember. I'm trying to show I do. "So we'd been away a while."

"Only a couple of days," he says.

"It just feels like forever," I say.

"You're playing it well." He strokes my hand.

I rub my eyes to blank out tears.

"After you went off to swim," he says, "I got up and went in for my circuit training. I was on my way when they called me."

"Have you told me what happened each time you've visited?"

"Yes," he says, "more or less."

"What about telling me more."

"A maid discovered you in a seizure at ten. You'd made it to our room," he's stroking my hand with his forefinger and leaning towards me with his head down, "at five minutes after ten, that is, you were still convulsing, half on the bed, half on the floor, when the housekeeper came in. They called me." He looks at his watch now. "I came in at ten-ten and you were still in the condition they call stasis."

"What happened next?" Do I really want to hear it?

"The ambulance came at ten-thirty. You were convulsing through around twelve. We rushed you in an ambulance over here. They decided it was best to sedate you. You were unconscious for a couple of days."

"Did I almost die?"

"Almost."

"That's interesting."

"Not to me," he says. If he isn't strong, he wouldn't be here. He could have taken off and I would never have known.

I think about all this for a minute, which is probably as long as I can hold onto it. "Did I pass out? Is that what happened?"

"Not exactly."

"Good. I'm sleepy."

Now I am awake. The world outside is light green. A visitor is here. "I'm confused about what's happened, but it seems best not to know." Tears come.

"It's all right." The man is very protective. I tell him I do know him. Just can't place him exactly. He sounds like Richard Burton; that's how I know the voice.

He hands me white roses. "Those are nice," I say.

"The flower of Yorkshire." He's testing me. This is supposed to mean something. Can I make this connection? "I'm your husband," he tells me. He lightly kisses me on the forehead, sits beside me and holds my hand.

"Did we go through this before?"

"Yes," he says, "we were at this spa—for our tenth wedding anniversary." He takes off his glasses and leans down to hold me. "I love you."

"Are the children scared? Did you tell them I'll be home soon?"

Now it's another morning. I am up early. I'm walking with a soft-armed woman in the garden. "It's wonderful to be outdoors."

"We walked by the rose garden yesterday."

"Oh, good, I'll have the surprise of seeing it all over again for the first time." And if we do something and I don't remember, then I can't miss it. It's the stretching, half a shadow, something I almost remember, that's a problem. Worse will be when it's a memory of a time I shared with someone and I don't have it, so in a way, they'll miss it, too. Solitary memories aren't as troubling. If you don't have them, you don't know.

My mind is still. I reach up to touch the branch of a tree. "This looks like a tree I love at home, a perfect tree, a jacaranda tree. Our family is known for being able to memorize easily—but I can't tell you your name. Today. Or his." I nod over in the direction of the visitor in white who's standing by the door to the garden. "And he's my favorite doctor."

She laughs, "That's your husband."

"Oh, really?" I laugh. "There's a surprise for you. Does he know I don't remember him?"

He kisses me on the forehead. He's immaculate in white jeans and shirt. He points out that he is my husband. He's brought me five peaches.

We sit in the hospital cafeteria, looking out at the garden.

He brings over plates, a knife for cutting the peaches, and some tea.

"We were at Champney's, a health spa. You'd been swimming. You got out of the pool and someone found you in a seizure."

Do I hear impatience?

"What kind of seizure? A heart attack? I'm too young for a heart attack."

"No," he holds my hand gently and firmly, "it seems to be epilepsy, but they'll want to do more tests later."

Epilepsy. Julius Caesar had it. Where did I get that? Why do I think I know that?

"Did you see it?" I know it's not attractive. When I was a kid, I had a friend, Carol Steinman, who had epilepsy. She was never allowed on overnights and went to very few birthday parties. Everyone said you wouldn't want to see her have a fit, see her foam at the mouth, her eyes roll back, she'd wet her pants. At Will Rogers's stables by the polo field at home in L.A., they shot horses who had fits. "I didn't know I had epilepsy."

I'm slicing a peach, carefully. It works better with my left hand. It's as if this hand and I greet each other like old war buddies.

He's watching my hands. "It might be something no one wanted to tell you."

"Maybe I forgot." Maybe the "friend" who had epilepsy was me.

He takes the knife from me.

"Don't worry," I say, "I won't kill myself over this."

He hands me a slice of one of the peaches. It's pale. "This is a wonderful taste."

I look at him. "You're seeing if I remember the flavor. I don't. But right now it's bright and soft, sweet and sharp, all at once, maybe more a fragrance than a taste."

"I'm not testing you," he says.

"Not really," I say.

"You're going home today." A woman rolls up the shade with a snap.

I try to picture what home might be, but that doesn't stop the scene from unfolding. I walk up the dark driveway to my house. I open the door. The house is dark and still and empty. I walk through each room and not only is nothing here, I know I don't even remember what was here. The kids are really gone. I can't ask—even if I do find someone to ask—because you can't say you don't know where your kids are.

"I don't think I'm going," I say. I don't want to go back. I live in this empty room with a book I've got to finish writing and a couple of snaps of my family. I can't remember the book. But I take it for granted I do this—like I'm not surprised I walk or eat or have had these kids. But they are gone. I mustn't let on that this bothers me. I pull on a long blue denim skirt. But I don't want to go back, so I take off the skirt. I can dress later.

"Don't you remember home?" She hands me the skirt. "You'll be wanting to put this on, luv. You'll remember when you see it." She looks out the window. Is she trying to remember a home for me? Or for herself?

"I don't remember. But it's been a while, hasn't it?" I'm not sure how long. A few weeks, I guess. My mind has become a pool. I put a fact in, an idea; like an image, it dives under and may come up as something else, transformed. Or not come up at all.

A man walks in. He's wearing a tight suit. Of course this is a doctor. Easy to see. They all look orderly. Except my father's heart doctor, who had asthma from smoking but couldn't stop. What do I need this for? Pieces of memory I do not need come through like dive-bombers dropping old flyers. I wave my arms at them, "off, off." The doctor's unsettled. "It's okay," I tell him, "I remembered something I don't need."

"Yes." He clears his throat. That's dealt with. "How are you doing today," he says to the chart at the foot of the bed, "still confused?"

"I'm kind of drowsy."

"Well, you will be," he looks at my chart, "you're on phenytoin. You were in status epilepticus when you were brought in to Mandeville. You're lucky to be alive." He has on a striped shirt and matching tie. He's the kind of doctor who regards illness as a disciplinary problem and likes prescribing things you hate.

"I spoke to my wife's sister," the Englishman says to the doctor. I've forgotten the Englishman is in the room. He's standing by the window. "She doesn't remember any mention of epilepsy and there aren't any records in medical offices in L.A. Jill had two convulsions during her sleep that I told our London doctor about last year. And she was given phenobarbital every night as a child, which Dr. Earl thought indicated epilepsy."

"L.A. doctors are largely interested in drugs," I explain to the Englishman. "Mandeville . . . but I live in Mandeville Canyon." I can see Jane Fonda riding her horse down the road dappled by the trees. What kind of trees?

"Mandeville is the name of the hospital, Jill."

"Of course it is," I say. So when did they put a hospital on the polo field? This is too confusing to argue about, so I just say that I used to use speed, but haven't had a drug or a drink for a long time.

"Really?" The doctor doesn't sound convinced.

The Englishman presses my hand and says, "She hasn't, not since 1969. When I noticed she had the seizures at night, our doctor sent her to a neurologist, who recommended that she go on an anticonvulsant. But," he looks at me, "she wouldn't."

"If that's the stuff I'm on now," I'm sharp about this, "I don't want it! I can't write on this—can't hold onto what I'm thinking, which means no writing, can't catch what I see." I feel drugged, handcuffed, struggling in an empty room, trying to fit a big jigsaw puzzle together.

"I shouldn't worry about the writing for now, my dear. You've been seriously ill," the doctor tells my husband, as if I've said nothing, "but you might want to confirm the situation with a new EEG when you return to London."

"But I want to go home."

The Englishman takes my hand. "London," he says, "is home."

I catch "is home." I hear his resignation. Not frustration, exactly. He seems to accept difficulty. That may be useful. It's not easy with a wife who doesn't have a grasp of your name. I look him over. I catch his character. I don't need to know where we are. Or where we're going today. He'll keep me safe.

2

*T*he dark red Jaguar has a steering wheel on the wrong side and a chauffeur, a notion that doesn't startle me.

"So, you have a Jaguar?" The Englishman is pleased that I recognize the car. But I'm from L.A. We may forget the name and the face, but never the car. The car suits him. I might not have named it so quickly if it had been a midlist car.

Mark, the driver, a tall man with the awkward charm of a shy six-year-old, is uneasy. Should he say he notices I don't know him or keep it to himself? He's English. He keeps it to himself.

"I'm sorry, I can't remember you. But I know you," I say to him. I hug him lightly, he blushes, and we are all reasonably more comfortable. I take my husband's hand (his name's on the tip of my tongue) as we sit together in the back seat.

"Mark's a writer," my husband says, "and he works as a chauffeur and gives us free rides."

"Until I write the best-seller, Madam," Mark says, "then we'll all have drivers."

"Then I'll drive," I say. Not while I'm on this stuff. I can't take that risk with the kids. Mark will help.

"I can't wait to see Johanna. And Jeremy. Yes. Are they scared?" I'm standing on a cliff suddenly with a flooding river of fears stretching out below. "You didn't tell them I have—this?"

"This is epilepsy, Jill—not another word you don't want to say. Johanna said she's seen you in what she called 'space-outs' when you've been overtired, so she wasn't surprised. And Jeremy wanted to get on a plane and rush over, but I told him that wasn't necessary."

I hear the midnight phones hollow in the dark. I'd rush to New York (from where?) to see my mother, to be with my father. I see my father, his coat flying behind him like the nun in *Madeleine*, rushing to my grand-mother.

Of course it is necessary. Jeremy shouldn't take a plane by himself. How far away are we?

I can tell by the way the Englishman talks about the children that these aren't his. This is another subject I'll drop. I also do not remember his. At least I'll see mine soon. I don't want anyone to know the scope of this thing until I've worked it out.

I have frustration enough for both of us. What do we grasp by instinct in a given moment, and what do we understand by memory? And what's the difference? He keeps his hand laid flat over mine as we're driven. I look at his profile, the dark blue eyes, long blond lashes. I don't remember how he became my husband, don't know anything about him or much about myself.

This is not my freeway. I've got an instinct for freeways. We're driving on another side. I'm not certain what other side, but it isn't what I'm used to. I have a feeling I'm off target here and it will be best to keep quiet. The man I'm riding with has enough on his mind. And the driver seems to know where we're going.

I feel a gathering chill, some nausea. I grab onto the door handle. The car swerves. It's not the car—it's me. I look at him. Maybe he can't see it. It must not show. I'm falling off the world.

"Are you all right?" He's scared.

No, I'm not. But I can't talk. I'm gripped. I don't like it. Want to get up out of it. I have a hunch, as my dad always said—about something—it's going to clutch me deeper if I fight it. I settle in, see where it goes; my God, like a sixties thing. A spinout. No, I mean the other word where I'm across the other room of my head, like a dance floor on its side. If it's an open tuna can and you're small and in it while it's spinning—and I'm off again the second I fall out, fall out, yes, that's the word, it's like fallout, say my mind blows up like a jigsaw, Dennis Hopper's paintings, ammies like lemon tusks, Billy Al's targets, atoms of images raining down. Out please, out-it's the bomb that becomes a fighter plane, hitting, then eases off, rocking away, like a shaky real old plane. It's weird, a kind of dizzy thing, but not exactly. Will the kids think I'm stoned? They have to know I wouldn't.

We are not on the freeway now. We are in the city. "I feel like I've been away for ages." What I really feel is, I've never been here.

"You may feel like you've never been here." The Englishman reads my mind. Easy read right now. A primer. No. A burnt-out library, all bare, charred shelves.

"No. It's just a little strange. How many weeks was I in the hospital?"

"Only a few days, it just seems long," he says. Seems even longer to him.

"This is Marylebone," he says. It's like Greenwich Village might have looked a hundred years ago. Small restaurants, useful little shops, images from Dickens and Rex Harrison movies. "It's below Regent's Park," he says. "You love to walk over there by the water and watch the swans."

But I hate to walk. Never mind. He doesn't need to know right now.

"We take friends to Shakespeare's plays here in the summer," he adds.

"We also do that at the Hollywood Bowl," I say.

"Probably," he says. Was that a scene from my life he was not in?—so, therefore, of little interest to him. This will be very depressing if everything I find to remember is before the Englishman came into my life. Nothing, and never forget this, existed before England got there and pulled it together.

There's really no scenery here. It's more like New York. I'm missing the mountains always on the edge of your eyes at home; to the left going up Sunset, to the right going down. Now we turn off and we're in the Upper East Side—hospitals, medical offices.

"This is Wimpole Street," he says.

"I can see the sign," I say, "and I've heard the name." I'm edgy and don't know why. "I know—Barrett and Browning." Tears. "I've lost my memory, not my brains." But how much of what we call brains is really memory? Can I be smart without it? How long will it stay away? Is it all erased, or is it in there somewhere under a dense new filing system?

The chauffeur turns off Wimpole Street and parks the car. I know I can't wait to see the kids, but my husband can. If he were really crazy about them, he'd have brought them to the airport. He picked me up from somewhere today. New York, probably. Or L.A. Does anyone go anywhere else?

"We're here." He's watching me. Do I know the house? It's large, red brick, two curved front bays with stone garlands, like a large-bosomed Victorian woman in a fancy bodice.

"But it's beautiful!" I exclaim. Clustered on the steps under the arc of a eucalyptus tree are pots of bright flowers, like storybook flowers, and juniper bushes like we have by my grandmother's grave at home. My father and I go there on her birthday and on Passover and Yom Kippur. After he says the prayers, he tells her what's been happening lately, how the movie reviews have been, what pictures he loves, and he tells her a joke or two in Yiddish. Then he breaks off a sprig of juniper between his forefinger and thumb and smells it as if he's catching her spirit. He puts the sprig in his pocket.

Before the Englishman has the key in the lock, a woman with thick, curly black hair and several earrings in her right ear springs out of the door and hugs me. She is plump, but very fast. When she greets me she has an easy Spanish accent, like L.A. Spanish. Maybe she came from California with me. "I'm okay," I tell her. I'm not. I don't want to go inside. She takes my suitcase from the driver. The Englishman's looking at the mail on a table in an entrance hall with a marble floor.

"This is our house," he explains. "The basement flats and the waiting room and two offices on the ground floor are rented; one to Fickling, the dentist, and the other to a sex therapist."

A young woman with silver eagle hair, maybe thirty, is standing in the foyer; she's not going to rush me. She's cool, elfin—sharp blue eyes—tiny perfect hands. She kisses me. She searches my face. "You don't know me? Is that bad? Or is that good?" she laughs, shrugs.

"All I remember is I'm crazy about you—you're not . . ." This could be my sister, but the age doesn't work, and she just isn't. "I'm really sorry."

She's holding my hands. "It's fine. I can tell you all my stories over again. I'm Laurie Lipton and I live in one of the apartments downstairs. I'm the artist. Gavin, the architect, rents the other flat."

"I wish the kids weren't at school today." Tears are starting.

The Englishman stops at the foot of the red carpeted stairs. "Jill," he puts one arm around me and one on the heavy, dark oak banister, "your children aren't here. Jeremy is thirty-four. I told you yesterday—and this morning. He's married and has a new baby, Phoebe. Johanna is married and lives in Connecticut and has a son, Justin. You can call them tonight."

"Why not now?"

"Jill," he says, "it's a different time zone."

Maybe I'll wait until I'm better, so they won't be uneasy. I hate when my mother isn't clear.

As he opens the door, I expect only the tasteful simplicity of the entry—never, never the wild wonder of this busy old library of color, of books and things.

I'm standing in the dining room of this house and I am amazed: paper masks and pictures all over the walls, plates with movie stars' pictures, antique toys perched on every surface, hedges of paper flowers tumbling down from the tops of bookcases, and swags of dark Liberty prints sweeping down over white lace curtains. It's not empty at all. When Johanna was born Jeremy went in and painted the walls of her room red, like his room. She belonged to his world and that was his color. My father's antique banks are everywhere—when did he give them to me? And the silver candlesticks. "These are my parents'," I say softly—and these big brass candelabra—Cary Grant bought them from my Aunt Lillian's shop. And the tall old art books, The Heritage Collection, children's classics. My brother is just beginning to read. He's the youngest. I reach up, "and these are my father's books, *Case History of a Movie* is terrific." I look at my hand; this is my mother's hand. This means my sister, Joy, and brother, Jeb, will be grown, too. This isn't memory. This is reason, which works slower than memory, like an old train. I think for a moment. Now I wonder if that process, my thinking, is to be trusted at all. There on the wall is a painting my mother did of me holding Jeremy when he was a baby. I look at my hand again. I don't want to look in a mirror. When I think too hard, I get dizzy. It's not a simple feeling. It spins off through visions I instantly forget.

"I've made some lunch," the Spanish woman with the curly black hair says.

"Don't you miss California?" I ask her. "It's all I remember."

"I've never been there," she says. She reminds me she is from Brazil. "I am Lilia, you see. You will be wonderful, you will see. I have ripe papayas. Oh, these are so perfect for the memory. You will tell me things about myself I have never dreamed. That is how good it will be."

The cool, young, silver eagle woman stands beside me and touches my arm lightly. "Do you know her?" She's looking at the large drawing of a serious, sharp-eyed woman. "She looks like Alan Rickman," I say.

"You haven't forgotten movie star names," my friend says.

"It's only my life I seem to be missing."

Sketched around the edge of a portrait is a strip of film with images of a child, writing. "I'm guessing the picture's me," I say, "which means I don't know her."

The Englishman is watching me. "This is Laurie," he puts his arm around her, "she's the artist. You may know her better."

"What I think I know about you may be what I know about my mother," I tell Laurie as we sit down at one of the tables. They're draped in my grandmother's shawls of wonderful old prints. "I'm confusing you, and being aware of the confusion doesn't fix it, although it's clear he's my husband," I say. I stroke the purple paisley and smooth out the toasty fringe that looks like peacock tails on one. "I've saved these for so long. My grandmother used to spread them out over her couch with the satin quilt my father gave her." I imagine the land in Russia looked just like this.

The Englishman goes into another room to make a call and I tell the artist, "I am blank one minute—then, the next instant, I'm here with you."

"Whoever I am." The artist looks at me with this sly glance she has.

"So," I ask her, "what does he do? How terrible I don't remember."

"He says he's a spy," Laurie smiles.

I think about it. "Could be." Could be I'm so caught up with my own work, I've never fixed on his enough for it to stay in my mind.

"He fixes businesses in trouble," Laurie says. "That's what he really does, but he works at home."

She's reading what's left of my mind. I'm confused again. I hate this about the kids being gone. I thought we were driving back and would be in L.A. any minute, winding up Mandeville Canyon, and there would be the kids, and the jacaranda trees. "I may be tired now," I say, and Laurie takes my hand in her own tiny hand. No wonder each line she draws is so precise, so smooth.

I listen to her carefully "So, we met in New York—of course, you sound like home."

"Yes, but we met here in London. You ran into me at a gallery and decided I was in trouble." Laurie explains. "I wanted to leave my Dutch husband. You weren't surprised I was leaving him. You said he wasn't difficult enough to wrestle through the time marriage takes to work. I would, I told you, but I have to live somewhere. I asked if I could rent the little studio flat in your basement and Stuart said, 'You can't rent it, but you can have it—you'll be here to keep Jill company when I'm on a business trip.'"

So he is like my father in this way, always generous. I wonder if, therefore, he is on the edge of a financial cliff. I wonder how often he goes traveling.

"I give you very little distraction," Laurie is telling me. "I'm an artist with a check once a month from a museum trustee, a scholar in Jerusalem. He pays me to reproduce ancient biblical illustrations. I have a brother in San Francisco and parents in New York. I can do a Bronx accent." This portrait she's done of me gives me the same sharp look she has. My mother also gives the people she paints her own expression. It's the true signature. On another wall in this room is a portrait my mother did of me with Jeremy when he was a baby. My eyes, his eyes, are sharp as her eyes.

"Right now, I'm a time traveler," I say. "I'm here from 1946 for a moment, then I stop by out of the fifties or sixties and seventies. I can't tell you a thing about the eighties."

"I think you can," Laurie says. "Come with me."

I go upstairs with Laurie. The children are probably in school, I figure as we climb the steps. There are four rooms on the upper floor.

First she shows me two small rooms like bedrooms. "These are your dressing rooms. They call them 'changing rooms' here," she says, "you can tell which one is Stuart's and which one is yours." That's easy. His is neat.

My changing room looks like a small used clothing store run by someone who doesn't want to sell you anything. The big ficus tree is decorated with ribbons, beads and chains. Hooks on the back of the door are draped with shoulder bags. There's no closet; the clothes just hang from a long chrome rack. A mirror standing on a pile of old *Vogue* magazines leans against the wall. Laurie burrows through a chest of drawers behind a basket piled with heavy sweaters. "Here," she digs out a big cosmetic bag, "from Clinique's 1987 free offer at Harvey Nicks, and these," she pulls shoulder pads from the cosmetic bag, "are all you need to know about the eighties. You loaned them to me every time I went out."

"Did I?" I'm distracted by a trio of chiffon gowns with hand-painted designs and beaded trim, black and red, white with pink and orange, like a fairy queen would wear, and a tawny one like a Navaho princess gown. "Zandra Rhodes." It's not fashion—it's ballet.

"Yes," Laurie says, "and Stuart bought the three of them in one day."

Then I hold up the lemon-yellow linen jacket

"Lagerfeld," she says, "you wore it to Ascot with a coral straw hat with a yellow ribbon." She perches pads on her shoulders.

"The age of the power knee," I say. "I don't remember the rest. Or the year. Don't tell me. I hate knowing how far off I am."

Laurie's slipping on high heels from the boxes under the rack of clothes. "You wear these to try to look like you're taller than I am. But it doesn't work because you can't walk in them. I'm taller."

"You aren't taller than anyone," I laugh.

"You forget. We went through this," she says. "He measured us. I am taller."

"I had a seizure. I grew. It's like a quake." Do all native Californians have this seismic tendency? Pull us too far out of our land for too long and we break up like this, shake into pieces. The swimming was too familiar, too close to home.

We move farther down the hall now. "This is your workroom." The room has dark blue walls, the color of my father's office at MGM, and is covered with posters of my father's plays, framed reviews and letters, a map of L.A. It's like a museum of MGM in the fifties with pictures of my father, the only writer ever to run a studio, with President Roosevelt, Adlai Stevenson, and Sam Goldwyn. And a picture from *The Wizard of Oz,* of Glinda trying to protect Dorothy from the Wicked Witch. The picture Sanford Roth took of Colette.

"My mind stops at 1954—no, here's 1960-something." I turn to Laurie, pleased. "Here's a picture of me with Barbra Streisand."

"You know her," she says.

"We were in a peace march together, I think."

"I think that was before you met," Laurie says. "I think she was in a movie with a peace march, and you saw it."

Maybe I can't tell what's real and what is a movie memory. Will any of my friends move forward in my memory, or are they all fixed here on these walls? Does film stay fixed in memory? Does it record on a different place then reality? Especially with the L.A. brain.

There's a long writing table, pitchers of colored pencils, a silver mug of black pens. Papers and books are piled everywhere. There's a picture on the desk of a little boy in a baseball cap. "That's Justin," Laurie says, "your grandson, Johanna's son."

"Right," I smile.

"You lie," she says, "you don't remember."

Laurie says this cheerful woman here with the tough look behind the smile is Johanna.

"But she just made this paperweight at school," I say, confused. It's a white ceramic globe with bluebirds set in terra cotta hands. What could it be, a year ago? It's as if I've come in at the end of a movie I really wanted to see. I will try to see how it went from what they tell me—I'll have to take it their way. Sometimes I won't be sure. But "not sure" is not the same as "I remember," not the same as "I was there." Their memories may not be my memories. But then I won't know, will I?

I do remember how I sighed when the phone rang and Johanna said, "It's your mother, Mom." Does Johanna really want to hear it's me?

Over the couch, draped with a gray, red, and blue afghan, there's a picture of dark redwoods and pines, like the forest near Stanford University, where I'd sit in the tall quiet, trying to find the focus to write. When my father couldn't write, he'd go for a drive in the canyons. We'd stop and climb up through the trees, not talk, or park and walk along the ocean on the way home. He'd talk about what he'd been thinking about by then. I learned from him that way; how to pull ideas into words.

I rub the fringe on the afghan between my fingers. "My grandmother made this," I tell Laurie.

I lift a picture in a silver frame from a long red table piled with manuscript pages. The picture's of a young man with a tough, tanned face, but there's a curve to his cheek. "He looks exactly like my father—with blond hair." He's standing next to a young woman and holding a baby.

"That's because he's your son, Jeremy," Laurie says gently.

I put the picture back next to the silver mug of pens. I touch the man's cheek, "I can't bear to say it . . . I don't know him. What do I do?"

"You can call. He'll put you on hold, but he'll know it's you."

I may not know him, but I know I never made his life easy.

"I think I'll wait."

"*I* have to tell you something terrible."

"Nothing could be much worse than that you don't remember me. We've done that. It's okay."

Now we're in the living room where my husband works. The walls are charcoal. It has his strong, unpretentious style. I am sitting in the big beige corduroy chair he hates (covering it with the Scottish plaid blanket hasn't helped).

"No," I say, "I can't remember how we met. How did I marry you?" I put my hand out. "You're handsome, I love you, and you're English. But you're not Jewish. How does my father deal with that?" What kind of person am I? I say "handsome" first.

"Jill—," he leans forward and holds my hands in his. He has to say something difficult.

"Stop," I tell him. "What time is it? I mean, what year?"

"It's 1990."

"You don't have to say it. My father's dead."

I can't remember if my mother has died, or when. I will call my sister. Yes, Joy, in L.A. My brother, much younger (maybe) is where?

"Did you know my father?"

"Yes, I met him a year before he died. When I proposed, he stood up. He was fond of ceremony, so I stood up and saluted. He took the bolo from around his neck and placed it around mine, and I promised to love and protect you as long as I shall live."

That feels heavier than just forever.

And he reminds me—most conversation for now is a reminder—"your last husband wasn't Jewish."

"I have enough to remember for now." I stop. "How many husbands?"

"Two. Your first husband was your children's father."

"I really can't find anything there. Not at all." I'm uneasy. "I'm opening a door to a room I don't want to look in."

"You will." He puts his arms around me close, but not clutching. His arms are perfect. I stroke them.

"I remember watching your forearms as you fixed something for your son to eat. So I thought you would be a good father."

The walls of the Englishman's room have agile pictures—are these Miró? And a stylish cartoon of him.

"That's terrific. Who did it?" I ask.

"My older son, Stuart Jr. He's an artist, a writer, a DJ, and he's part of an ambulance crew in Florida."

"So that's all he does?" I look at the picture of Stuart Jr. I can feel the same urgent presence.

I wait until Stuart is asleep tonight and find my children's phone numbers. It will be easier to talk to Johanna than to Jeremy. Daughters, like women friends, enjoy the phone.

"We already talked this afternoon," she says, then quickly adds, "I'm glad you called again. I'd be foggy too if I'd been out as long as you were."

"Did I call Jeremy this afternoon, too?"

"Yes," she says, "and he's really scared. He says you talked to him as if he's a stranger, and had forgotten Grandfather was dead. I told him you'd be fine in a couple of weeks."

"Thanks," I say. "Do we believe that?"

"Absolutely," Johanna says.

We weren't part of a social group in London the way I was in New York. There, friends would be calling, reassuring, dismayed—maybe disbelieving until they spent time with me. Stuart sometimes refers to himself as a troubleshooter. Now we are like a pair of troubleshooters, tackling this problem alone together.

"I told you we were coming," Barbra said, arriving one night for dinner with Donna Karan. I made spaghetti fast. She was baffled, intrigued. How is it that I have not forgotten how to cook? Maybe it's along the same lines as not forgetting a song.

I forgot we were invited to one couple's house for dinner. "Losing your memory is no excuse," the hostess shouted. (Who was she?) "Everyone forgets things. You must keep better records."

I cried so.

Stuart's extraordinary journals of our life together since we met have helped me sort these years out, which doesn't mean I don't fight over things I don't actually remember but don't much care for. Some things just don't feel like me, and I hate some things he says I said. So I say I didn't. Which changes nothing.

Also, he does not mention what I am wearing in most of his journals. And that makes it very difficult. I spend as much time thinking about what I'm going to wear as I do thinking about what to make if you're coming for dinner. These notes add dimension to memory and sometimes they'll place a season, but we fight over what he's added in later.

"No one," I said, when I was looking over his notes for winter '95, "no one was wearing a red tweed Valentino then—not even me."

"But I gave it to you and you loved it."

"Not after '88 for sure." How can I tell him that? It's a gift. I must have it somewhere. I'll wear it. Sometime soon.

In one of his journals for 1984, he says we had fettuccini, poussin, and bread pudding for dinner. He was probably hungry when he wrote that.

"There was no poussin," I'm guessing this part. The eighties are fairly blank, "and I wouldn't have creamy fettuccini with creamy bread pudding. You just wish we'd had bread pudding. So, how did we meet?"

"You'd heard me talk in Westport."

He was talking to a group of people about how he stopped drinking. We both believe that most great spiritual programs do best, like restaurants and movie stars, when they hold onto their mystique. So I won't reveal things about our program because it works, and always has, along its own lines of protocol. One of the principles is you don't talk about it. There's no problem, though, for us to tell our own stories.

"You told all your friends you'd seen this Englishman. One of them told you everyone in Westport knew you'd seen me—and that my best friends were jazz musicians and that I only went out with models. Your friends Marcia and Dolph knew me. They told you that if you thought you'd had trouble—you didn't know trouble."

"So, I was warned—which only made you more appealing."

"You did very well. You found out everywhere I went, everyone I knew. You sat behind me at a lecture by Anatole Broyard and at one of Gerry Mulligan's concerts. You tapped me on the shoulder and, in a very low whisper, asked for a cigarette, which I handed you. I didn't have to look over my shoulder to see who it was. And I knew you didn't really smoke."

"So you were intrigued . . . "

"I'd heard you were smart, a good writer with a rough story. You seemed harried."

Research journalism is excellent training. I can figure out what I might have been wearing, what I might have been working on. But I can research the thrill I had when I saw his dark green Jeep Wagoneer pull into the parking lot in front of the Sherwood Diner in Westport? Yes. Can I measure it up to the catch I feel when I see him coming down the street or walking into my workroom, and hold the sense of it, edgy as it remains? Does he know me? Will he talk to me this time?

"Did you know that was my red Jeep you were parking next to?"

"No, but it was always next to mine when I'd come out after having coffee with friends. I noticed you sometimes sitting at the window wearing Laura Ashley smocks. You were reading *Middlemarch*."

I was probably trying to look more English than cowboy. "You should have read that by now," Anatole told me.

"Most of the time you were writing."

"Someone told me you worked with Hugh Hefner on a few projects, and I knew that meant models."

"Probably," he says gently.

Your memory is in trouble if, among other things, you repeat the same stories more than once a day. Or is it more than once a week, or less than once a day? Forgetting these memory scales does not worry me.

What interests me is why I remember some things and not others. I remember the acid-green glass candlesticks are from Joanne Segel, but I'm not sure for which wedding. Gil and Joanne gave me my second wedding, but I couldn't tell you when it was. But I do remember everything we ate at Johanna's house this last fall and every word Jeremy has said to me this year. But then, that's because after I talk to him, I write it all down.

I can tell you exactly how to pitch the look I shot across the diner that night I met the Englishman. He had picked up a coffee to go and had just turned and started toward the door.

I put everything I had in that look—didn't spare a second of energy for "Is this a good idea?" I was saying with this look, "I am designed for you. I am everything you need."

"You came across the diner and stood by the table. I have no idea what I was wearing."

"That shows it doesn't matter after all, doesn't it?"

"Yes, but I remember you were wearing a khaki shirt and khaki pants." His big tortoiseshell glasses were perfect for the heavy blond brows and the silky wave across his forehead.

"I asked you if you would care to come and have tea sometime, and you said 'Right now would be perfect.' We forgot about the coffee I was holding. You followed me to my house. Our cars, as you pointed out, were perfect for each other."

"I may forget your name, but I won't forget your car."

We are sitting across from each other on low, soft chairs. The lights are low, too. We're holding tea cups in our hands, as we carefully place each line we say, setting words and thoughts out in front of each other like offerings. Do my responses show I understand you? Is this interesting enough?

I can remember Stuart's house, the house where I went with him that first night. I think it's the American neurologist Barry Gordon who says it helps to arrange words in categories. Houses lead directly to flowers for me.

"There were white lilies on the piano," I tell him. "You were playing a jazz song I'd knew I'd like from then on—but I don't know what it was."

"The song was called 'It Never Entered My Mind,'" he says. "I talked to you first about writing—wanting to write. You said something Martin Luther King, Jr. said, about believing you're everything fantastic you want to be. Trusting your excellence. You remembered things that well then," he goes on, "and you said you'd always loved the idea of going to England. Your favorite teachers and school friends had been English. But you'd never been out of the United States—you'd been working all the time and raising your kids, always into a deadline or a school term. And—this was closer to it I guessed—you said you hated going anywhere you couldn't drive."

"It's a California thing," I say now. "So, did you ever get me out of the United States?" I smile. I love the line of his jaw. Am I really married to this man?

"Where do you think we are now?" He leans forward.

I close my eyes, shake my head. "Well, I guess . . ." and look around the room. Do logic—this is not Connecticut. "Upper West Side?"

"Jill," he says, "we're in London."

"Well, it's been a long day." We hold hands. "Let's stay where we were in the story—about how we began—how you got me here."

"I did say one day I'd take you to London. 'We'll have a Jaguar waiting for us at the airport,' I said, 'and I'll drive you up north to Yorkshire, through some of the most beautiful land you'll ever see.' You said, when you saw it, that it reminded you of chaparral on the Santa Monica Mountains. We stood on the Scarborough cliffs, watching the northern lights. Then, later, we put the car on a ship to France and drove down to the Côte d'Azur, speeding around the Mediterranean."

"You made it sound wonderful—and what I probably thought was, 'Oh, sure.'"

"I wanted you to believe me, but I knew it would take time. When I put on some music, you said, 'You're a nice man. I don't want to throw you,' and I told you I hadn't read your book, didn't know the details, didn't want to know yet. I knew you had some tough, dark years. No one winds up with our determination to change just for a bit of heavy drinking."

"What if I'm somewhere one day and I forget I don't drink?"

"You won't forget. It's something like being left-handed; you know it. One of the most important things you told me then was you hadn't given up on the idea of loving someone. You don't forget that. We had that in common. A lot of people wind up loners, can't imagine finding someone to be with without drinking. You'd been hurt, so I was careful just to talk easily to you," he's holding my hand, "to let you know me, trust me."

He's doing the same thing now, circling in. I can feel that rise you get at the beginning of the titles in a movie you're really going to like. He reminds me about writers I know and love, about this writer's march we went on. Artists we're friends with, concerts we've gone to, how I hated classical music when we met.

"Do I really like it now, or do I tell you that?"

He's like a professor. He has big, clean, and perfect hands. I'm looking his hands over. I put them up, press them close to me and say so quietly, "It's new to be falling for someone you're already married to." And at the same time I'm thinking, *I'm really going to learn a lot here. It may have been this way the first night we met.* "So what did we do next?"

Is he too sweet to talk about sex?

"I could see you liked your reputation as this raunchy authority. But I didn't think you could easily take off your clothes with the lights on; that without drinks, without something, you were shy. I told you that after the hallucinations I had at the end of my drinking, I was scared of the dark."

"How do we sleep if I can't sleep with the lights on and you can't sleep with them off?"

"I told you we weren't at that part yet," he touches my face, "and we aren't."

"So where did we go from there?"

"And you turned away and told me to ignore what you write about sex. You said you were really bad at it, hated it."

"And then . . . ?"

"I figured out you had this thing about trust. The next night we went to a meeting, then I took you to hear John Mehegan play jazz. We all came back to my house, talked until late about the blacklisting in Hollywood. You didn't feel they understood your father's role. They were older, had a different perspective, but you made it clear you had actually been there in the middle and that you never had a child's perspective, but an observer's. I liked you for standing your ground. You are challenging, interesting."

And I'm thinking he's aware and smart. Like my friends, like Sandy, Lynn, and Josie, I have an interesting husband.

"Where are you?" he asks me.

"I'm thinking of all of us together having dinner at Lynn's in New York. Anatole and Sandy, Lynn and Dick, and Josie and Peter."

"Jill," Stuart says, "Anatole Broyard is dead. Lynn Nesbit and Dick Gilman are divorced. And I never knew Josie Davis. She has been dead since 1973."

Will I remember that she is gone the next time I think of Josie?

Johanna Mankiewicz Davis. Every generation has a few mythic losses, potential snapped off, people who were going to be everything. In L.A., Josie had been ours.

Stuart and I are in the living room on Wimpole Street again. He's telling me about the second night we met. "You came back to my house. My son was there."

"You were working on something together, right?"

"The stereo," he says. "After Philip, my youngest son—he was staying with me—went to bed, we began to talk again. Then around dawn, I put on Billy Joel's 'Just the Way You Are,'" he tells me, "and we began to dance."

"Did we?"

He's put on some music now. He dances easy and firm. He lets you move but he's there with you, isn't going to let you go, but isn't holding you back. "I like the way you dance." He holds me tighter. Now he will remind me how it is in bed.

That's what he means. Can he talk about sex? Can I?

But where's the anxiety, that catch that tells you this is love? He feels too safe. He feels like a protector.

Is that sexy?

We are sitting on the edge of the bed. He kisses me. He kisses like movie stars kissed on screen.

Did I kiss movie stars?

I remember as I kiss Stuart that I knew, as I watched him that second night with Philip, he was exactly what I wanted and what my children needed. Solid paternal values. Strong, tawny forearms. Firm hands guiding Philip's hands. Philip, a fairer, slighter version of Stuart. Then, yes, his oldest son, Stuart, Jr., an artist, I think. His daughter, Susan, went to school with Johanna.

"Do our kids get along?"

He stops where we are. "They don't really know each other and they live far apart, but they understand their differences."

Would it have stopped me falling in love if they hadn't?

I put myself first, while I told myself I was putting the kids first. I look at him in the soft light here. He's putting on another record—no—tape— but that's not it, either. Whatever it is, I like how he moves. But then, my kids would have had no interest in the accent or the blond hair. They might not have wanted another father. They might well have preferred to keep things as they were.

"The next morning we slashed our wrists, mixed blood—this was before AIDS—and swore to be together for the rest of our lives. And because you said you tended not to pay attention or to hear only what you want to hear, you swore to listen. I gave you my oath to love and protect you."

A protector is sexy. Even though I couldn't remember what he said, I felt it in his presence. More an animal instinct, the way you do fall in love before there's anything to remember.

"And what did I swear?"

"To always hear me, to listen to what I'm saying."

"Do I do that?"

"Most of the time. Yes, you do."

"Do you hear me?"

"Yes, I think I do."

About two weeks after we'd taken oaths, we sat across from the minister, Ted Hoskins, in his office on the church's second floor. A tall, robust, silver-haired man with a giant presence, he'd be a minister, a whaler, or a sculptor. The sun always lights up behind him.

"But I'm Jewish," I said, "and Stuart's Christian. Would you marry us?"

"There's only one God, isn't there?" he said. "You'll be fine as long as you keep your priorities clear—as long as you have the same priorities." That sounds very simple.

We agreed we'd be no use if we drank again, so that was our first priority. We were each other's second priority. Everything else must come next.

"Children, parents, work?" For me the work had always come first. "My children have resented me more for that than for anything else. Including, I think, drugs and bad men." You can't put a mother down if she puts you first. "I was just beginning to be able to give them everything, to put them first, to give them a home." Maybe it was too late. But I wanted to keep trying.

"Jill, they're twenty-five and twenty-three years old. Maybe you're giving them an opportunity to explore their own lives," Ted suggested. He was putting it in a nicer way.

"I don't agree," I said, "but I want to be married to Stuart."

A year later, we'd married, flown to England, and were standing on the balcony of the round room in the tower of the Holbeck Hotel in Scarborough looking at the northern lights.

You can fear, when you look back and see where it's all come out, that you didn't do the best thing. But you can also understand how you felt then and know you would do the same thing all over again. This is a remarkable level of pathology. Or acceptance.

I come down at 7:15 the next morning. The Englishman is sitting at the kitchen table having blueberries on corn flakes.

Even with my back to him I can tell he's sad. I'm having a test in an hour at the Churchill Clinic. He's probably scared he'll have to take care of me forever.

I woke up early today, read about him in one of his early journals. He had a younger sister whom he had to carry around during the war. And when Stuart—or his mother?—had TB, he lived with his great-grandmother. She ran the movie theater in her town, so he saw movies free, any time he liked, and practiced love scenes coached by older girls.

This is the test: MRI—Magnetic Resonance Imaging. Dr. Rudolph arranges to have me attached to a kind of a sled, my head in a cage, and slid into a steel rocket through a world of gauges, buzzes, hums, taps and quirks. Far below I hear Stuart singing me back through time with the Lambeth Walk, Glenn Miller, and at least twelve verses of "Green Grow the Rushes." Was that a Yorkshire song, or was it a movie with Richard Burton, who sounds just like my husband? Has the same sort of lower lip.

Josie and I followed Richard Burton around Palm Springs one day. Joe Mankiewicz, her uncle who was directing Burton in a movie, had told Josie that Burton was sleeping with Elizabeth Taylor. He'd tell Josie anything, and the look he gave Josie was the look we wanted from Burton. Older men, especially directors, were sexy. They would make good coaches. They know how to light you, move you, and charm you so your face lights up. They'd know how to touch. We wouldn't have wanted our parents to know. The whole point was to move out a bit to test new territory. The men knew how far to go. Some people's parents looked the other way.

Josie hasn't called in ages. I'm scared for an instant. Hate this. Then I remember I don't remember things. Maybe she has called and I've forgotten. It is better to slip that away to deal with another time. I'm not sure why, and by now I don't know what it was I was thinking.

"You see, I'm not going to die," I smile, "there's no brain tumor."

Dr. Rudolph explains that the epilepsy pattern has been managed by the phenytoin, and the blood test shows I can handle a higher dosage if it becomes critical.

"You know, I used to be given phenobarbitol every night as a kid," I say. I understood the sniffs of powdered ephedrine were for the asthma, which made you crayon brilliantly. But phenobarbitol is not for asthma. "Could that have been for epilepsy?"

"That was a customary way to manage epilepsy then."

I always figured I could get around a condition, get outside of it, and have this "Happy Ending" sort of life, but I'm always dragged back to face an item of inventory I've tried to escape. And now, here's epilepsy.

I remember passing out as a kid and being removed from phys ed, which was blamed on the asthma, a far more acceptable condition. You could stop breathing and die. But you didn't wet your pants, foam at the mouth, or have your eyes roll up into your head. None of which worked as an acceptable attention grabber at Warner LeRoy's birthday parties. You'd practice lines.

Jean and Susan Stein's parties had a classy form, like New Yorkers trying to do English aristocracy. Linda and Warner's mom, Doris Vidor, understood L.A. style: fast dialogue and great art.

"Sleep, rest, lower stress, and normal living," Dr. Rudolph taps the results into order, "and your confidence should return as much as possible."

"Sleep, rest, and shopping," Stuart says as we leave the office quietly. What he means is we should take a walk, which along with tea is the British solution to most disagreeable situations. I have not forgotten that I hate to walk or that he has learned to call it shopping.

We cross Oxford Street, which looks like Wilshire at Fairfax. I'm jarred by the faces, the images. He catches that, takes my arm. "You like it when we get to Bond Street," he reminds me.

It's late morning, and the clouds float by animating the figures in the windows. "I think I'm glad I never knew I had this," I tell him.

"Maybe it's part of your imagination . . ."

I interrupt, "Don't even begin to tell me it's okay."

He pulls me back fast from a crossing. "You have to look the other way!"

"I know!"

"You don't—not now. Listen to what I'm saying. I've known since we met that there are fundamental differences in our ways of thinking."

"Exactly. I'm L.A. Female, Jewish. You are completely British Male, Christian."

"What I mean—just hear what I'm saying."

"Bond Street windows—perfect for my concentration. I'm listening."

"You see, I usually work my way through to a conclusion in a rough sort of rational way," he continues. "You almost always jump to a conclusion."

"Immediately," I say. "That's true." I put my hand on his arm, "Sorry."

"I thought you did this because of your 'feminine intuition' or 'artistic insight,' but that's not the complete answer."

"But it is partly."

"Partly, but I now believe that you 'jump' to conclusions the way a person jumps onto a departing bus. You need a resolution while you can still hold onto the topic under consideration. Most of my thoughts are buses parked, waiting for my selection. Your thoughts are transient and have to be jumped on before they leave you stranded."

"By the time I get to it, it may be gone."

"Something like that," he says.

"Nothing could be a bigger gift," I turn him away from the Ralph Lauren window so that he's facing me, "than to have someone think about how your thinking works."

Sleep, shopping, and sex. I hold his hand.

5

On the shelves and around pictures on the walls of my workroom there are bronze plaques and medallions, humanitarian, film, and theater awards given to my father. A picture of Roosevelt and posters of my father's plays. A sketch of Ethel Barrymore when she was young. I am halfway in my father's room, halfway in my own. I am sitting in the rocking chair, wrapped in the gray, red, and blue afghan my grandmother crocheted for him. I remember turning the wool into balls as she held the skeins up around her hands.

My father was the only writer who got to run a studio, which some thought was like putting a fox in charge of the hens. It was, in the sense that to write you have to put yourself outside so you can see the whole picture. To run a studio well, you have to believe you *are* the whole picture—you have to be everywhere at once.

I used to listen to my father talking to writers when they'd come to watch rushes on a picture that needed more footage, or when they'd call him late at night troubled over rewrites. I'd like to call him now.

I thought I'd put L.A. behind me when I went East to write books, but I guess I catch my voice from my roots, where the rhythm's as easy as the sound of surf rolling back and forth. I hitch up my memory of surf, to get it right. When was the last time I saw surf? How does the movement go?

I look at the date on an award to my father from the Anti-Defamation League. I remember my father's birthday. I try to write down the numbers on a pad of paper from the table nearby with a pen. It's dried out. I take another.

I hear the Englishman's footsteps. He is stopping by here on his way to bed. I catch the Aramis drifting by and the violet-striped Turnbull & Asser robe. Bond Street and its territory are fixed in memory's hard disc.

"I need new pens," I say. He is standing in the doorway. I look at the dates and up at Shaw. "My father's dead, isn't he?"

"Yes. He's been dead for ten years."

"Our family doesn't live much past seventy, that's how I figured it out. He'd have been eighty-five. And you would have said something about calling him, you know."

"He died ten years ago."

"Did we go through this before?"

"Yes."

"I wonder if I remember less if it's something I don't want to know?"

"Of course. It's called denial."

"Was I there when he died?"

"No, you weren't. We were on a plane flying back from our honeymoon. There was a message at the airport from Jeremy."

Jeremy is my son. Now I see it. Jeb is my brother. He is fixed in my mind at fifteen, when we'd drive around town in my new red and white Mercury. Jeb is my father's son. And now, Justin is Johanna's son. A lot of J's. This may make it easier. I don't have to race around the alphabet to catch names. Stay with J's and I'll run into someone I love.

I look Shaw over. "Were you there when your parents died?"

"It wouldn't have made much difference if I was, would it? We went up to Yorkshire every time my mother had a heart attack and it didn't make a difference. She died because that was her time." This, I seem to remember, is the Yorkshire approach.

"It makes a difference to the one who's sick," I say. "It made a difference to me that you were there, even when I didn't know." Which is my family's approach. Break an ankle, and we'll all rush to you on a plane. With Petak's Russian coffee cake. I can tell you the name of the coffee cake, but at this moment here, I'm using my husband's last name.

"I'll be there in a minute," I tell him. He has an elegance in the soft hallway light.

I cannot remember how sex goes with him. More to the point, I don't remember ever being very easy about it.

I am in my changing room. I'm looking at this stack of slender orange Hermés boxes outlined in chocolate, in each one a scarf he's given me. You'd think they'd date them; Hermés dates its narrow brown cotton ribbons, which I have hanging here on branches of my tree. If I'd kept each ribbon with its box, I'd know if this pale blue Christopher Columbus one came before the one with Navajo chieftains. I don't like wearing them so much as I like having them, feeling them. I wore this first one with the gold keys on our honeymoon, wondering how his family in Yorkshire would feel with me so grand and having all these things. I had this scarf on when we rushed from the airport right to Seventieth Street when my father died. My mother touched the scarf when I kissed her. "At last, someone took her to Europe," she told Stuart. Hermés scarves— deductible, perhaps, as basic mnemonic aids.

This may be my favorite, fields of wheat sparked with poppies, corn-flowers—and here—the rabbit with the scarlet eyes. I wore it with the two-piece St. Laurent cotton dress—mistake to give the top away when shoulder pads came in. My daughter is called Rabbit. She's fast, swimming or skating. I only wear the red and royal blue Eiffel Tower or the navy zodiac scarves on really outgoing days. I didn't think the French were into astrology. It is convenient that Stuart has the same sign as my mother and my children; an Aries never lets you forget its sign.

I put on the blue and white Lanz nightie patched with bits from the strawberry print that Josie and I bought for old times' sake when she came back to L.A. in the sixties after her mother's first heart attack.

The Rafelsons had just moved in right below me in Laurel Canyon, near Brooke Hayward and Dennis Hopper's house. Josie and I were at Toby and Bob Rafelson's on the Monday night of the big '66 fire. Bob and Jack Nicholson were playing poker with Roman Polanski, who looked at us all like we'd never been anywhere. And I hadn't then. I remember the sound of sirens lacing through the night around Rafelson's and Nicholson's moody dark laughter. I can feel the tension now. Is this the way memory defines itself? It wasn't just the regular summer heat. We'd all been on the gallery walk-around, drinking and catting our way along La Cienega. I was feeling sleek that night. I had hipbones and I'd been on TV with Mort Sahl twice that month, talking fast enough so everything sounded sharp. We were doing our basic down politics riffs, when Brooke called in shreds, taking turns on the phone with Josie and me. Then Dennis came over in a rage. He'd put these pictures out on the lawn to dry, "and Brooke just went out there, man, and turned on the sprinklers. I'm not fucking kidding," he said. Rafelson was already figuring how he'd use this in a screenplay and Toby got Dennis a bowl of something. Because of speed, I don't remember sixties' food, but it was probably chili. Like sex, we did not talk about food then. We just made a lot of it and shared it all around with anyone who might drop in.

In Daniel Schacter's wonderful book, *Searching for Memory*, he talks about the memory wars. Some memory explorers believe once it's gone that's it. The connectors are burned out. Others go for Proust's idea, that the scenes are in the storage vaults, waiting for a "retrieval cue."

I remember this night with Brooke because I hear a siren. I left the Rafelsons—it went something like this, I think—and I brought Brooke and her kids, Jeff, Willy, and Marin, home with me that night.

The fire engine I hear now in London calls up the image of our children lying together on the shaggy royal blue carpet, on Merimekko cushions, the moon coming in the huge window which reminded me of my mother's art studio.

We were hoping they thought this was just a fun overnight. Or is that what I'd like to believe we were thinking then. Did we see their sensibilities as people so clearly before they grew up to tell us?

The siren wails off beyond Marylebone. I'm in our bedroom. Shaw's pulled down the covers. I slip in beside him in my nightgown and hold on tight, patting his shoulder. He kisses me with this big, long Gable close-up kiss.

"Could we turn the lights off, I wonder?"

"Of course," he says.

"It's still light in here," I say. "Streetlamps?"

"Maybe the moon," he says.

"I know you've seen me without clothes on," I say. "I'm sure."

"That's true," he says, "but you're not easy about it, so that's okay." He's even more charming when he's amused because it doesn't happen easily.

"I can't remember you laughing." I look at him lying back against the dark paisley pillowcase (Ralph Lauren, late eighties.) "I can't remember—do you have a sense of humor?"

"You told me yesterday you didn't think so." He's stroking all along the curve of my body, lightly, carefully. "And you said that was fine with you."

"I guess it's not okay—but it's how it is."

"Hush," he says, and his lips are cool and perfect. *Stay,* I tell my mind. *Stay here. In the present.* It slips, like silk, like a scarf. I then feel it starting, that gathering feeling, the arising, rolling, buzz.

"I can't . . . ," I say, "I'm scared I'm having one of these things." Petits mals. That's what they're called. It's a couture condition. Hermés could do a petit mal scarf. I can give them images. If I could catch them, tie them down so they stay in place. "Your hair looks silver blue in the moonlight." I touch his brows.

"How many people," he says, "can begin again? I think it's interesting."

"But you remember where we were before."

"That," he says, "is what makes it so interesting. Do you remember when we first met you told me you couldn't come unless you were tied up to the bedposts? You had a lover who made gift wrap. He'd tie you up in his newest ribbons. But I didn't have bedposts and wouldn't have tied you up anyway. You talked a great game then, the way you wrote, and I knew the charm of it was I had a newcomer."

"So to speak," I say. I lie quietly as he turns towards me and puts his wide strong arm across my body like an honorary ribbon, a banner, the kind Princess Diana had on with badges when I sat next to her at that concert. I don't remember the concert, but I remember her easy, open attention that broke the ice she assumed you'd feel sitting next to her. "Isn't this great?" she whispered at one point, touching me with her arm, the kind of arm that gives a real embrace, sturdy as Stuart's arm on the other side of me.

"What are you thinking about?" he asks. He's now got his leg across me so my thighs stay this way, open.

The wind is coming up from behind the full moon, whipping round and clearing bits of sky so the moon looks set in it like a rough-cut antique diamond. I used to sit at my window when I was a kid. I looked out over the drive winding down between hills covered in English ivy and I imagined I was in an English castle. Then I'd stand at my three-way mirror, sideways, so I'd look over my shoulder the way movie stars do, lifting my eyebrows, trying out smoky glances as I brushed my hair (make it stay blonde, God, let me be glamorous, please). If my parents were out, I'd go down to my mother's maid Dorothy's room. She gave me bourbon. We'd listen to her Bessie Smith jazz records and she'd call me "Glamour."

One night the Ku Klux Klan, in their white pointed hoods, burned a cross on the ivy-covered hill right by the big stone reindeer. This was because black people were not supposed to live in Brentwood.

When I first met the Englishman, he actually had a six-foot-four-inch woman from Jamaica living with him, towering over him. She kept him warm when he got the shakes getting sober. They were a forceful combination, but she left the day after we met. She left her clothes cut up in shreds all across his lawn.

"I can't remember what I'm thinking about," I said, "what do you think about?"

"You," he says.

I doubt it. But it's a perfect thing to say.

He's like a large, endlessly passionate animal; curvy, dense flesh you can disappear under. I'm a ship and he's the sea. Then we turn, rearrange legs and I'm the sea and he's the ship. Can he keep my attention on him, on the present, instead of this kaleidoscope inside my head dense with tough images.

Stay here. Consider how the tongue of the wave curls back down and sucks up the flat spill lying up there on the sand, and what is the word you have for that; it curves up there, never the way you expect it, always getting the edge of your beach towel.

His hand feels like sun.

Set your mind here, on a stone balcony in the village of Èze in the South of France. I'm lying naked in the sun. Like warm silk. Drape it just so it shows this part, like the Eiffel Tower part on the Parisian scarf—and then, with just a twist of your neck or a touch of the wind or his arm around your shoulder, and the ivory triangle comes out like a glimpse of my mother's nightgown. Cut. Start again.

The power builds; the energy's on. It gets still, focused and silent.

"Stop! Don't!" I'm silent, can't talk through the seizure just as you can't talk when you're coming, not really. I can't tell him it's happening. Is it an orgasm or a seizure? I've got to get out, don't tie me into it or I'll go all the way and never come back. I get out from his hand, his leg, and lie here panting, crying against his body. I don't want it to be just sex with him.

Just Sex, she's saying, and Sex is so insulted, having such an open field day in matters of life and death and politics and out of its closet altogether is Sex in this century—the nerve, Sex says, and it is a matter of nerves, I have to remember to ask the neurologist whether in fact the seizure and the orgasm are related, and if it is a kind of come.

I try to turn to loving him, but he's withholding here. "Not yet," he says. Giving sex, "I want to see yours but I won't show you mine," is the oldest child trick. You want to see a grownup so you don't have to spend quite so much time wondering what they (especially the directors) look like without their clothes on. Do you? Want to know, really? Yes. But that's different. Is that adult abuse? Thinking how all your parents' friends, especially the movie stars, look without their clothes on? And what they do, exactly. Especially the Democrats.

The villains, Robert Ryan and Richard Widmark, would have interesting games. They would tie me up and take off one piece of clothing for every line I got wrong. Do they know this is what I am thinking at deli dinners on Sunday nights? Lana Turner went upstairs to hug my brother good night, and he said he could feel her real breast against his face.

I wonder if I saw more, knew more, and, as I think my father did, only let myself feel what I could handle. This I can't deal with. "No, no, no," I twist away, "I can't go there yet."

Find something lovely, like the peonies on the scarf he bought in the south of France. Folds and curves, edged in crimson, dark as it is here, where he's trying to find our connection, carefully, he's winding his way along; a sure tread he has with his fingertips, like a seasoned hiker here, chivvying me along, the way he did down that French mountainside, or was it Yorkshire, or a trail down from Mulholland where I'd sit and watch Lupita Kohner and Maria Cooper taking turns riding Traveller, the school horse, waiting for the weekend when my father would be home and take us for a hike along the trails through the chaparral off Tigertail Road.

My father was easygoing on the weekends; his walk, his voice, the expression on his face changed. He became Jimmy Stewart. We'd take turns building up a story, scene by scene, as we drove up the canyon.

This is not useful for coming. Go red, sharp, dark. Cut it.

Turn on my side away from him, shaking. He's a wonder, says it's okay, hush, and holds me close, rocking me in his firm, steady arms.

6

*T*oday he takes me down the street to Sagne's, a bakery with a pla-
toon of marzipan animals and cakes like white chocolate bouquets
in the window. "Princess Alexandra used to come in here every year with
her lady-in-waiting to place her order for Easter eggs," Stuart tells me.

"How nice to have someone whose job it is to go shopping with you,"
I say.

"You do," he says. "When we first moved here, you thought no one
knew you. Then one day, when you'd left a cake you'd bought, someone
from the shop came by our house with it on her way home. It had a note
on it, 'American lady writer—Paid.'"

We sit down for a quick cappuccino. Clement Freud is sitting across
from me, glaring down at his cream puff. A trim, sporty woman with a
great smile hugs me, reintroduces herself quickly, "I'm Jacquey Visick."

"Does everyone in town know?" I ask Stuart after she's left.

"She's a journalist," he says, "one of your favorite friends. Calls every
day to see how you are."

We pick out croissants, almond and chocolate, and a raisin Danish. I
ask for some bagels. "You don't like the bagels," Stuart says.

"Maybe I will now. You'll see."

A young woman is in the downstairs waiting room when we come back. She has a little boy with her, maybe four years old, cheeky and restless, sucking on a candy. "Don't even begin to tell me you don't know me," she says, "don't even start."

"Of course," I laugh, "I know you." I do. But not her name. Or his name.

She has curly, expensively tortoised hair tangled into a barrette. Her eyes are round and glittering dark. "It's a good game," she says, "you may get away with fucking murder with anyone else." She tosses aside the *Vogue* she's picked up. She looks at a woman you know is waiting for Fickling the dentist because she's reading *Country Life*. "Yes, I said 'fucking' in front of a kid in broad daylight."

As we go upstairs, she reminds me, "I'm Judith Marcus. I own a company that makes commercials."

She's watching me, concerned. "I thought it was one of your dramas, but you really don't know me." The memory loss is real. She jokes it away. "If you don't remember anything, I can tell you anything and you'll keep it safe. How convenient."

"Can I get the paints?" Noah, her little boy, asks.

"Sure," I hesitate, "if you know where they are."

"You let kids paint on the walls upstairs," Judith says.

We go into the kitchen.

"And you say I'm best," Noah says.

"You are." I hand him a glass of water and some of the brushes from the mug on the kitchen table. Will I remember anything she tells me, and if I don't will I be able to fake it?

"Cesario, my lover, is charming," she explains, "but his days may be numbered. He lives too far away. With our work, his wife and kid, and Noah, it's very hard to make time to see each other."

"So he's married?" I ask.

"Of course. Why can't they be wonderful and single?" she asks.

"They can, but then they'd be less interesting," I say. I can see she gets bored easily; she runs a studio, has a child and a lover and an ex-husband. She surfs through her days like a man does with TV channels.

"I'd always find something wrong so that I can go back to Cesario," Judith says, her face in the refrigerator. "He is perfect, just available enough for entertainment." She peeks around at me. "Not enough to get in the way of a marriage."

"You could always see this analyst downstairs, you know. That's what he does. Works out marriages. Stuart says he's meant to be very good."

"You're saying 'meant to' like that to sound English," she says. "Stop it. His name's Mendell. He's a sex therapist. That's not the problem." She puts her arm around my shoulder.

"So, do you have any idea what happened to you?" Judith sounds tough, but she's far more sensitive and easily hurt than Laurie, who has the self-absorption and assurance of an artist who knows she's a genius. "Can I put out some cold chicken?" Judith asks.

"You see," I say, as I get a blue glass from the shelf, "we had an assignment to write about Champney's, the health spa."

"I know," Judith says, "I told you to do it, that you'd love it. I didn't tell you to get carried away."

"Literally." I tell her about the anger I had when I was swimming. And suddenly I remember why I was angry. "He took chocolate bars with him. And he napped instead of exercising. I guess I got wild with rage."

"I'm about to do that," Judith is putting pieces of roast chicken on a platter. "Cesario's wife thinks they should move back to Italy. How can she do that to me?"

"You know, the very worst thing can turn out to be the beginning of something amazing. Or did I tell you that already?"

"Probably," Judith says, "at least you don't act like you're tired of hearing all this."

"I'm not tired of listening," I say, "but I'd be interested anyway. Even if I hadn't forgotten."

The Englishman has walked into the kitchen and is pouring another cup of coffee.

"Convenient you should turn up now."

"Memory does not matter to her," he puts his arm around me, "she invents everything anyway. I was in an exercise class when they called me and told me she was having a full-blown seizure."

"Convulsion, everything?" Judith lifts her eyebrows.

"It was so bad, he won't tell me what it was like in the ambulance."

"If I were you, I wouldn't want to know."

Noah comes back in for another piece of chocolate croissant. Judith says no. I hand it to him when she gets him a glass of milk. "He knew you'd do that," she says. "So you were swimming?"

"I'd made a mile—or nineteen laps. I remember the sign that it takes twenty laps to swim a mile. So I dove in. The water was perfect, easy water, turquoise pool water. I worked hard from the twelfth lap. By the time I got to fifteen, I knew I'd get there, do the mile, even though it was hard going. I was counting the laps then, turning at the end of each one, to see how far I'd come. After eighteen, I pushed on into that fever you get when you've wrapped something big. It was only a mile—but for me—"

"For anyone," Judith says. "A mile for any of us is a big deal."

"Well, I did it. And lost the rest of the week—and in a way, everything else. I don't know what I don't know until it comes up, and then I realize I don't remember it."

"When Stuart said I was coming over," Judith tosses back her hair, "did you have an image of me in your head?"

"Of course," I say.

"Not at all." She looks me over carefully. She can case you fast as Walt Winchell, as Lenny Lyons. Judith's shrewd and tough on top, full of longing, generous underneath. I wonder if Jeremy is like this.

We set out the gray plates with the clipper ship pattern at the small round table in the dining room. Judith and Stuart are talking about her studio, but when I join them the subject changes to Cesario. This is not unfamiliar.

"I'm really trying to end it." She takes a piece of chicken from Stuart's plate. "He doesn't like white meat," she reminds me.

"Can you just decide to end something, just like that? Give yourself time—otherwise you'll feel terrible when it doesn't work," I say, "like quitting anything."

"I think you can just end it." Stuart is disagreeing with me. "You end with the action. The feelings go in their own time."

"Like you with Peggy?" I snap. Peggy was the girl he stabbed when she went out on him. This was quite some time before we met.

"You remember that?" He's encouraged. He explains to Judith how it was every day when he went to see me—how I'd see him in a different way—regard him with a different expression.

"I'll send you Oliver Sacks's book about the man who mistook his wife for a hat," Judith says.

"Maybe I thought he was a Jaguar, but I wouldn't remember.

Sometimes I hear him play the piano when he's not even home," I say. I haven't told him I've been looking in his journal, trying to catch up on our story.

I am not entirely certain if this is the same day as the last piece of time. Stuart has put postcards on the kitchen wall from holidays we've had together, dinners with Ted and Vada Stanley, Bruce and Lueza Gelb. This interests him even more than the pictures of his own children and grandchildren—which we are also taping on the wall next to the kitchen table so I can start to understand, to hold onto who they are. I have lists of their birthdays and their phone numbers and addresses, which I've matched with the pictures. I also have pictures of my children when they were little, when I seem to remember them best, and pictures of them at other stages. Judith is here and there are croissants, so it may be the same day. Or this may be a regular thing we do.

Judith and I are looking at pictures of my children taken when they were with us during one of the summers Ted and Vada loaned us their beach house. Here we're sailing with Bruce and Lueza. Lueza's standing on the bow of the ship.

And here I'm standing with Barbra. We're looking at each other in exactly the same way as my father and Sam Goldwyn are looking at each other in this other picture over here.

"My father may be younger than I am in this picture."

"He may be the same age Jeremy is now," Judith points out.

"And I don't know Jeremy as well. Is this typical? That we don't know our children as well as our parents?"

"We probably don't let our parents know us," says Judith.

"Probably. I haven't seen him for ages," I say.

"You saw him last Thanksgiving," Judith says, "you were just getting yourself together. Jeremy brought Phoebe with him."

I don't remember, so I move on to the next picture. "This is Justin." He's got a Yankees baseball cap on. "Here's his mother, Johanna." I get misty.

"Don't get that way," Judith says, "Johanna has no patience with that."

"You mean you don't," I say, but then neither do I. "I look at these riveting adults and I don't quite understand who they are."

"How would you?" Judith says. "They've changed. You see Jeremy and Johanna maybe once every year for a few days. How can you expect to know them?"

Is she saying I don't see my children enough?

"I haven't forgotten guilt," I snap at her. "I really don't need you to tell me that I haven't seen them much." I'm too sharp. Lighten up.

"Look," she says, "this isn't memory. We're all different to our parents. If we lived next door, it wouldn't be any easier—maybe harder. Denial is the traditional way to handle family relations."

"Even if we were in America, you wouldn't remember the last time you saw your children," Stuart says quickly.

"You're trying to show me that it's not all your fault, that it's not only because we are living here in England." I'm snappy. That's because Judith's here and they'll go in his room as soon as she can and talk about her work. "Maybe," I tell him, "you're really wondering how much I remember about our own last trip. Do I remember the cliff overlooking the ocean, the tiny donkeys on the beach, and starfish in tide pools?"

"That's so sweet," Judith says, "you do remember."

"Does it matter if I don't know exactly what we saw? It's like splicing a few frames of film to put the story together." I say this to Judith, to remind her that I know about film. I have the defensive determination you get when you're older to remind newer people that you were around in the business, whatever business, long before they were. You do it when they don't include you, because they're trying to show that they've arrived and they want your total appreciation.

But then my defense is usually light-years behind the time, as I probably am. Do they splice? Call it film anymore?

"She's talking about Scarborough." My husband gives Judith another piece of his croissant.

"She knows," I say, and so he knows he's not off the hook, "of course, if we were at home, my kids could come visit and show me who they are. There would be accidental times when we'd run into each other and have ice cream, and be able to make up for the last fight."

"Spontaneity won't come easily," Judith says, "with Johanna in Connecticut and Jeremy in L.A."

Now Laurie's tapping on the door. She says she's come up to ask if we think the landlord will care if she paints her kitchen.

"Why would you do a picture of a kitchen?" I wonder.

"I don't want to paint a picture of the kitchen," she says. "I want to paint the kitchen walls."

"Since when are you cooking?" Judith looks her over.

"I'm not. But if I liked the color of the kitchen, I might."

"So what color is it?" I come into the middle. This is in case we miss daughters who don't get along.

"You shouldn't see the color until you're better; in fact, it could set you back years," Laurie says.

"Years I can handle. I'm very good on 1958. It's yesterday that's a problem."

As my husband puts more chicken on Judith's plate, he tells Laurie, "Mark's looking at the other flat across from you."

I tear out a bite of bagel. I want a flavor to go with how it looks. This is like seeing a movie without the score. "Bagels don't taste like I remember."

"That's because you don't really remember how they taste," Judith says, "and you invent dissatisfaction to distract us."

"I'm not being dissatisfied . . . "—don't let her hurt you—"I'm just puzzled."

"That's possible," Stuart says. Judith splits the chocolate croissant and hands half to him.

Sound sharp, I tell myself. "So who is Mark?" I turn to Laurie. "And what color is the kitchen?"

"Mark is the writer who pretends he's your driver, something I wouldn't expect you to forget," Laurie answers, "and the kitchen is chartreuse. Doesn't that ignite the memory?"

"It's probably easier to remember things you hate," Judith eats.

Not sure. I toss up a set of flash cards, like mug shots, on my mind's screen—former husbands.

"I've thought of showing her movies she hates," the Englishman says, "but there aren't any."

"That's not true," I say, "I hate most movies with Dan Duryea." I'm delighted. "You can't say my memory's all gone if I remember Dan Duryea."

"That's so useful," Judith says.

"Well, it is," Laurie says. "We could put her on a game show."

"I remember chartreuse," I say, "you'd wear it with cocoa, maybe a turban hat—or with fuschia jersey. On its way out, it ended up on leisure suits, and I remember violet, my grandmother's shawls." I brush the crumbs off my favorite placemat, which I put out this morning—"this and the other one with ranunculas."

"But that's not your grandmother's," Judith says, "you bought those all on sale at Liberty."

"I don't think so." I'm dizzy with the confusion I get when I'm wrong.

"It doesn't matter," Laurie says quickly, "they look the same."

"The important thing," Judith says, "is you haven't forgotten the things that matter." She wants to get down to business now. "I need to talk," she tells Stuart, and leads him into his room.

"I could come down and paint your room with you," I say to Laurie.

"That's a perfect idea," she says. But you don't have to remember to sense how someone is about being alone.

Judith pours a cup of coffee for Stuart, throwing in two sweeteners and some milk. I wouldn't have remembered exactly how he likes his coffee. She pulls off a bit of the almond croissant on his plate as she hands him the coffee. I hope she's in love with someone.

Noah wants me to see what he's painted. I take his hand and he leads me upstairs to the wall by my writing table.

This is a new morning. I look at a blond male; imagine a Henry Moore sculpture you can sleep with. There's a tree outside the window. The man is no more or less familiar than the tree; we are no more or less connected. I go outside the room. I'd like to have some people around. When I'm on my own like this, I feel like I'm on a windy roof. Nothing to hold onto. I'm in the changing room looking at a rack of clothes. I have a lot of chino clothes. I find some pants, a shirt, a jacket. Arthur Miller was dressed like this when we marched down Fifth Avenue in New York. I can't remember what it was about. Here's a picture of my father with President Kennedy. I guess I'm a Democrat. But I can't remember who's President now. Are my values and political attitudes different now? What if Stuart's not antiwar and thinks nuclear testing is a good idea? He's English. Is he fond of the Queen? Will things I used to like to eat be distasteful? Will I love music I used to hate? How much of choice is influenced by association? Will everything be a brand-new choice? Are these khaki clothes comfortable because they bring back a moment I liked? And they are so worn in, it may be more than one moment. I sit down on a needlepoint cushion to pull on socks. Who did this needlepoint? There's a bag of shoulder pads on the floor—for making clothes?

Something. Shoulder falsies, for affectations of height and power. Did I run a company? God help it. How many of these questions did I ask yesterday, and how many have already been answered?

I go into a long mind skid. *Take a ride,* I tell myself.

I've got a Mustang convertible. It will be in the garage. I'll drive out to Santa Monica Canyon to see Sherman and Joan. Sherman Yellen's out here working on his new play.

I go out the front door. Wrong. You're not home. This is certainly not L.A. I walk up this street. There's bound to be a garage somewhere. You can always tell by the car whose house it is. Judith has a Jeep; shows you can run a family and work at the same time. You feel like a World War II general.

I stand at the corner, confused. But this could be Wilshire, around Hancock Park. On a Sunday. I know I'm wrong, but I can't make the connection.

"Are you lost?" a sturdy woman rumbles up along the curb on a motorcycle. Is she picking me up? She has on a red and turquoise Nike outfit, bright helmet; she's a giant toy. "Jill, it's Prue." She takes off the helmet and blonde hair tumbles out.

"Of course. Hi!" I laugh. "I'm fine." I have no idea who she is.

"It's okay," she says. "I haven't seen you in ages."

She's saying that to make up for my embarrassment at not knowing her. The English can't bear embarrassment. And Americans are a major source of embarrassment. Look what I do to this authoritative corner of substantial Victorian architecture, standing here, looking uneasy, having the queasy appearance of an L.A. cliff after a 6.5. She throws a chain around the front wheel of the cycle and padlocks it to the fence of this house just down the street. I stand under the arc of a eucalyptus tree and pinch a bit of the bush to my right. It smells like home. I knew what it was—but now it's gone. She unlocks the old paneled door. This is my house. Don't be crazy. Don't be that crazy, please. We walk in. I look in my pockets for a key to our flat.

No key.

I'm there five minutes when Prue says, "We've got a patient soon, but you can stay in the waiting room."

A small woman with thick silver and black hair comes by, she does an instant freeze frame with her eyes. "I'll call Sagne's and tell Stuart you're here. He's been looking for you." She has hair like my Aunt Lillian. But her hands and her jawline are fragile, exactly like my mother's.

"I've only just gone out," I say.

"No," she looks at the biker, "no, you haven't." She picks up a phone on the table in the black and white marble hall. "This is Laurie Lipton," she knows her name carries weight, "will you tell Stuart Shaw that his wife is back at home." Then she looks at the biker. Prue. Yes, Prue.

"How are you doing?" the biker says to Laurie, who shrugs.

"It's good I live in the basement because I can jump out the window and it won't kill anyone."

"A friend of mine, an opera singer," Prue says, "would really like you to do his portrait."

"I'm open to proposals," Laurie says.

After the biker—yes, Prue's her name—goes into the dentist's office where she works, Laurie gives me tea in the kitchen, which has only one chartreuse wall left. You've been in England over a year when you start serving tea. After five years it's what you really want to drink.

"I'm thinking about this biker I met," Laurie says.

Prue reminds me. "A travel photographer. We might be fine for each other."

"You're never crazy about the one you'd be fine for," I tell her.

"You are," she says.

"Possibly. But maybe we decided we'd save everyone else from the trouble we are and made up our minds to be crazy about each other. It was best of all for the kids. We're here. Out of their way."

"You don't mean that," she says.

"I don't remember if I do. It feels familiar." I know I don't mean it "I might not have a choice." I can change the subject easily and just say I don't remember what we were talking about. "So, you could bike all over—to America even. And you'd have free transportation all over. Fast."

"That's the part you'd like. You're terrible at matchmaking. Besides, I work in black and white. I need someone who dresses in black and white. I think I've met her. She works in the City—smart, funny. Conservative. Ideal for me."

"But can I trust you—to know?" I ask. "Is she also remote and unavailable?"

"So, your memory returns. She will bring me to a new level of despondency."

"That's important if you want it to work." I look over the walls. "That chartreuse is terrible. Thank God you're changing it."

"I told you. I never lie." The black and white drawings are everywhere, big wide-open expressions. The woman eating an apple, which is actually the head of a child. You look and you hear the large voices; you don't have to know them to know them.

"You draw like I wanted my mother to paint," I tell her. "You catch the edge. I used to watch my mother paint gentle, ravishing images filled with love. Then at night, after we'd had our glasses of sherry, she'd tell me the true stories, the savage stories. Why is negative more real than positive?"

"It isn't," Laurie says, "except when I'm in a depression. You don't hold onto depressions well, but you're very tolerant of other people's depressions." She tells me that when she came back after her mother died and had hidden out, I'd left bowls of chicken soup by her door. "You never tried to cheer me up."

"I wouldn't think of it." I look around the neat rooms.

Sheets of paper lean against every wall; each one covered with real people, their reality raised by her fine pencil lines to a rich new level of grotesque precision by the perfect details of the hands, the eyes, grimaces—surreal characters you know are someone's relatives.

Do I write about my family like this? Do I remember my family? From hello, I miss you, to goodbye, I love you, I know more about a person I met an hour ago.

"So why did you paint this wall chartreuse? Pop art?"

"I didn't do it," she says, "you did it because Stuart's grandsons were visiting."

"Really?"

"You took them to Paris after a couple of weeks here," she says. "Nathan, he's the oldest. He's very smart—extremely handsome. And Kenneth, who's original, funny. Kenneth may be schizophrenic. He was diagnosed before you went away. You were talking to your friend Vada about him."

"I don't want to remember this part," I tell Laurie. I look around at the kitchen. "And I hate this green."

"I told you so."

"You never tell someone 'I told you so.'"

"My family always does," Laurie says. "It expands guilt to a remarkable degree. You can't plead ignorance."

"Yes, but in L.A. we have no guilt. We have earthquakes, fires, and total destruction, and get to start all over again every couple of years. How can I have guilt without memory?" I ask.

By not calling my mother, of course. So I call, but it's been so long, her phone has been disconnected. This could be to show me how hurt she is. Or because she is gone. The loss of memory is refined denial.

I don't want to ask him. He's playing Brubeck on the piano—"Take Five."

He has that dreamy, pleased expression on his face when he plays, an expression men have only when they're playing jazz or driving new cars, pastimes that require little memory, only composure.

My mother will see this loss of memory as a new act of aggression; having lost my memory, I may have lost most of her life. For who has more of her memories than I—from old Hollywood love stories to the first edition of *Lady Chatterley's Lover* she gave me when I said I wanted to learn about sex—"It's all here," she said. But you can't see the loss of memory. It's the kind of thing someone like me, who craves attention, designs as an affectation. I can also design ways to cover it up, masks and games and distractions (if I can remember I've decided this). No one really needs to know about it.

Stuart takes the memory lag more seriously than anyone because he's lost as much in it as I have. He's kept his journals every day since we met. I go over them to catch what I can so the next moment we are talking or lying together, I can say, as a lover would, "This reminds me of that morning in Èze village at The Chevre d'Or."

Now, I have read this last night and cannot remember the name of the place. So even when I read the journals, I can't hang onto what he's written long enough to use it. Then, too, if I read it and say, "That was wonderful"—whatever it was—he knows I am not seeing the memory, not catching the colors again; the sound of those waves breaking on that shore, or smelling the roses on that one hot day in London in Regents Park. *A hot day in London,* I say to myself, *can that be true?*

I remember so little of who he is or what he does. But he also knows better than I do what I let just drop and what I really can't hold onto.

While Stuart's playing music, I am going through old papers in my changing room. Well-named room; it is here that I become who I think I am today. I have found a note in a child's hand from my sister, Joy, when she was ten and I was twelve. She's apologizing for antagonizing me.

I cannot remember if I've called about my mother. Why should it be surprising, that the hard things to deal with are the most difficult to remember?

"It's been ages since I've seen you," I tell my sister, "so what are you doing with yourself?"

"I'm seeing my clients." She's catching me up.

"Clients?"

"I'm not in real estate, if you're wondering. You know that." She reminds me she is a psychotherapist. "How long do you think it's been since we've seen each other?" she asks. Her voice has the analyst's seductive concern.

"About . . ." Silence. Throw a number into the air. "Five years?" There's a question mark there. "You've been alone in L.A. with our mother," I won't say "my mother."

"Yes." She's not going to pull me along here. "It's okay to say," Joy says, "that you think she's dead. She is. You sent her packages in the last weeks that she loved—some drawings. She thought you were very dear." My sister uses "dear" the way my mother and father do. Or, rather, used to do. I have to say "hot" or "cool." Never "sweet."

"She lost her memory, too—for a different reason—but she wouldn't have known you had been here moments after you had spoken to her," she adds, "so it didn't matter that you weren't here."

Joy is telling me I hadn't wanted to come. I can imagine (if not exactly remember) that she got the hard part. There wasn't that much of our mother to have around, and during the years when I was there, I tore off more than my share.

*D*r. Rudolph sent the test results to Christopher Earl, a neurologist who will interpret the brain scan. I have three attacks on the way to see Dr. Earl. He's large, with white hair; probably belongs to the Lords' Cricket Club. I must send one of their cream flannel cricket uniforms to Justin.

Dr. Earl is looking over Rudolf's report like a bank manager going over exactly why your checks have bounced. "You have a focal abnormality in the left anterior temporal region, which indicates temporal lobe epilepsy," Earl says. "You were mortally ill." He's concerned I might not be taking this seriously enough. He also points out, "The attacks can be brought on by thinking about it or talking about it."

Chris Earl has sent a note to our doctor, Peter Wheeler, who has an office on Sloane Street, which has so many great store windows, it is possible to forget you're going to see a doctor. I think I'm on Fifty-Seventh Street until I'm jarred by a New York accent going by with her husband. You don't notice American accents at home, which is how I can tell I'm here in London. The same way a Londoner would suddenly notice an English accent on Madison Avenue.

Our doctor is tall, stylish, and handsome and stands up to greet us.

"Hello, Peter," Stuart says, and I wonder why Brooke didn't call to tell me they were in town.

Peter shows me his letter from Chris Earl. They write to each other using their first names. Chris is an excellent Christian name. Definitive, one might say. They probably meet for lunch, have port and one of those plates of everything horrid you cut off a pig. Chris has written Peter, "One could make out a good case for her taking regular anticonvulsants and I'd suggest Phenytoin, which you originally prescribed."

We leave the office and walk up Sloane Street.

Let's say I have now gathered together some things about my husband which I can lay out, as I do my scarves, in my attempt to say I know him, or know what he likes enough not to bother starting wars by suggesting things to do that he hates.

He doesn't ride the Underground because he's claustrophobic, which is also why he does not do elevators. He doesn't water plants or cook. And he does not eat anything that is not chocolate, or ice cream, or has not been slaughtered or hooked. He's easily bored, and does not believe what is new is automatically interesting. After reading this over a lot, I remember what's missing: he doesn't go to Hampstead. So I don't go. I might think I'm in Laurel Canyon and never come back.

"You think if we walk on Bond Street and Sloane Street, we'll catch the fever for making money," I say to him. "The shopkeepers nod at you, like you're trooping your colors." I take his arm. He always keeps one clean credit card to refresh the self-esteem during dark days.

We stop at L'Express and stand on the steps, looking down, waiting for one of the tables along the wall in the front. You don't need a memory to walk into a restaurant and know the hot tables. This is a sixth sense, or maybe what Daniel Schacter calls "priming." Like you learn your own mother's voice before you're born, you learn what's a good table. To play it safe, never take the first table you're given at a restaurant where they don't know you.

We sit down and survey the menus. I tell Stuart I'm confused and angry that none of the doctors want to discuss the issue of my missing memory. "Nowhere in the letter did Chris Earl say that he liked me, that he understood why I don't want to be drugged."

"Uninvolved medical remains inconceivable to you, doesn't it?" Stuart says.

In L.A., doctors always want to be your best friend, which our family always thought was a good sign, because doctors were very smart and wouldn't be friends if you weren't interesting—or going to live successfully enough for some time, so there would be social benefits. Friends made midnight visits, and free script, and I'm using "script" here in the pharmaceutical sense, not talking about movie scenarios, although some doctors didn't mind putting in a word here or there about a hospital scene—and maybe getting a mention on the credit crawl.

"A doctor should be your friend," I tell Stuart, "especially if he can't fix you." There's an irritating couple of women at one of the smoking tables in back, each on her own phone. You hear them over everyone, making dates with someone else.

This is the kind of place Brooke and Josie and I would go to at home. I order a fruit punch. They have paper tablecloths. I look at the menu. "I'll have a Caesar salad."

He's surprised. "You usually have a Niçoise."

"I decided to try the Caesar." I can't remember the Niçoise. I take out the pen from the—what's the word for it?—in my terrorist jacket—we bought them in chino and white at Ralphie's when we first came to London. They have dozens of pockets and these fabric racks that look like bullet slots, where I keep pens. "I think Brooke Hayward and I sat here the last time she was in town. That was the time a woman came over and told me that one of my books changed her life, and Brooke wondered who the hell she thought I was. I was hurt, but I'd love it if Brooke walked in today. I miss people I actually know." I'm drawing a purple coyote on the tablecloth. "I haven't done this in years," I say.

"Why would you think Brooke would be here?" he asks, but he's more interested in a tall Nigerian model who has walked in with a man out of Hemingway wearing a rumpled white linen suit and a cigar.

"Because," I said, "we just saw Peter—"

"Jill," he thinks a moment, "Peter Wheeler is not Peter Duchin, Brooke's husband, although he'd be pleased by the confusion. They have a resemblance, and you were having lunch with Brooke in New York when you saw that woman. It was Esther Ferguson. Aside from that, it's fine."

I'm not asking who Esther Ferguson is. "I fucking hate this," I say.

"Are we talking about mixing people up or drawing on tables?" Stuart asks.

"Mixing people. I like drawing. I make a drawing of Stuart dancing with the Nigerian model. Drawing is easy now, not writing. It used to be the other way around.

People say we have a happy marriage. This is because if you ask each of us how we do that we say we do it "his way" or "her way." This, I am seeing, is true in the sense we have figured out what matters most, and we each give in that territory. Except in these two places. First, Stuart cannot watch any one TV channel at a time. He says people who do watch one channel at a time have one-track minds. And second, I do not comprehend how you can love family effectively from a great distance, as he seems to do.

We are quiet, watching stylish colonials in black outfits have lunch with their phones.

"And another thing I've forgotten," I say over my second fruit punch, "is my concept of God."

"You have several concepts," he says, "you change them like clothes, which may have been what brought the subject up. Your concepts all fit. You should read Freud's *Civilization and Its Discontents*. It would help you understand where your beliefs come from."

"They're all from movies, don't give me such credit."

"Don't give yourself so little," he says. "I haven't got patience for that today."

"I'll give you a break," I say as we leave L'Express, "let's walk, but I want to walk home a different way." I'd kick pebbles if there were any to kick. "Let's go down that street," I point to one, "I don't remember it."

"We've done it ten times," he says, "there's nothing to look at; you'll hate it."

"Like that movie I wanted to see which I can't remember—and you said we'd seen it a week ago and I didn't like it! And I'm sure we never did see it. But not sure enough to be absolutely sure." I'm standing still, hands on hips. "This is not fair! You asked me where I wanted to go, and I said. And you don't want to. So we don't."

"The reason I walk you on the same streets is exactly so you will remember them!"

"Could you once do something not for me?"

"Easily," he says.

And we walk home fast on streets with no shop windows, not talking.

He glances at me, and I glance back at him. He must feel that our lives will be empty now—made even darker with my blank presence.

I'm not sure I can retrain my mind to remember, let alone to write books.

9

*T*oday he is walking me over to see one more doctor, an expert on epilepsy. It's busy downstairs. The dentist who rents an office on our main floor is just coming in. Fickling is a tidy little man in a pistachio doctor's tunic. "Back from the holidays?" he says as he shuffles through the mail and gives my husband a quick smile. "Nice day?" By saying "Nice day" to me he means "Are you okay." They all know. In England you express concern or convey friendliness through references to weather or holidays.

"Lovely day for a walk," Fickling adds. "Lovely day" is when it's not pouring. I hate walking.

This epilepsy doctor has asthma. "Ventolin's great," I tell him as he takes a hit from the blue inhaler. Maybe it's magic to have a doctor with the same trouble you have. I don't think so.

"I can't tell you what it was like without it," he says.

Maybe it's not magic. You want a doctor who doesn't get sick, to show you that it's possible. "I can tell you," I say. "Not breathing is what it was like." I also want this visit to be about what's wrong with me. "Could what's happened to my brain be damage from not being able to breathe? Or drugs? I used to take speed a lot."

"You took what?" He's embarrassed. He's not had an addict in his office.

"It was a long, long time ago," I say, "and it started with the Ephedrine. Remember when you could only get it in little packets of bitter white powder? You'd sit there unable to breathe, waiting for the packets to be opened. I was sure I was going to die."

"You can remember that?" He's suspicious; am I in here for some kind of fix? Or am I certain everyone's suspicious of me?

"The memories are coming back in patches," I say. "I've got bits of this. But L.A. in 1944 is not particularly helpful if you're living in London in 1980-something," I snap.

"Especially," he says as he writes something down, "when it's 1992. This has no connection with drugs." He looks at one of my reports again. "I'm afraid it's not that simple. You have always had epilepsy."

"I don't think so," I say.

"It may never have been active, or you may not remember."

"Someone would remember," I say. I remember Carol Steinman had epilepsy. No one wanted to be alone with her. She was never allowed to be with us. You were warned when you did see her not to upset her. She was like special crystal, she just might fall off and shatter into terrifying, shaking pieces.

"At least it will stop now," I say. "The pills will control it."

"They will modify it. But," he inhaled his Ventolin again—he shouldn't use it so much, but that's something he doesn't want to hear from me. And besides, this is a hard thing to tell me—"it won't stop anything." He shifts up in his chair and leans over his desk, elbows on it, hands folded. "Your husband told me you were a writer." He slips a few pages of my records out of a folder.

"I could have told you that," I say. "I remember. Could I see your notes?"

"Of course," he says.

There are ten letters. How nice. One calls me a sixty-four-year-old authoress. Like ten years older than I am. Another one says I'm fifty-six. And here I'm fifty-one. Do I hear another offer? I'll go for forty-nine. And my birthdate's on each one; but then with all the work it takes to learn anatomy and drugs, it's understandable you skipped subtraction. I'm no one to talk, and here we have all different visions and scan results.

"Jill." Stuart puts his hand on my shoulder, I shrug it off.

"Mainly, I want to know if I'll be able to write again."

"I'd like to reassure you," he says, "but I'm afraid that kind of effort will be very difficult."

"Are you saying it's impossible?" I ask.

"I wouldn't make it that important in your life."

Snap my fingers. So it goes. "You're saying I won't be a writer again." I want to hear it, like you want to see the enemy.

He's sad. "I wouldn't think so. First of all, most patients find that any mental stress triggers attacks. Even minor attacks are debilitating and affect the memory. So the more you work at writing, the more stress and the more likely the memory's affected. It'll be very difficult to remember what you're writing. You'll be able to do social correspondence, that sort of thing, of course, but I shouldn't plan any larger projects. You're going to have to remain on medication."

"Maybe there'll be a new medication," I say hopefully. "You never know."

I've always put writing first, scattering my life and the lives of everyone I've loved, everyone I've known, and now it's been blasted away.

It wasn't just me who was hit. It's a restless Sunday afternoon. Stuart's driving me around the city. I try to look as if I've seen it before. He tries to look as if this is something he feels like doing. "It's like taking a dog for a walk," I say, "isn't it?"

"I'm trapped behind invisible bars," he says, "honor, duty, loyalty— and nothing I give you helps."

"I can't help you fix me," I'm crying now, "and I can't fix me—so maybe you just go. Put me somewhere, and go." I can't think where he could put me. "I don't want to be like my mother, where you have to visit." Or his mother. But I never want him to know I'm not crazy about visiting his mother. "This isn't what I want for you, or for my children. Maybe this has happened to show me. Touché. So just take me to the ocean, throw me in."

"London doesn't have an ocean."

"Well, how the hell did you get to America!"

"The city is built on the Thames, you know that."

"I never heard of a city without an ocean."

"Chicago. You like Chicago when Studs puts you on his show."

We're at a bridge over the river. "Let's go to the other side."

"It's the Valley," he says, "you really don't want to go."

"I want to do something else, something different!" Which we both understand is crazy, since just about everything feels different and new. He stops the car and we are on the other side of the Thames, looking out at Big Ben. "I wish I were looking at Manhattan and coming home," I say.

He sighs, impatient.

"Your writing will come back," Stuart says firmly. "Never mind what they say—it will come along in steps." I used to listen to my father talk over story outlines with his writers when he was running the B unit at Metro. They'd cobble together pieces of ideas and characters and things to say, while my mother painted on the screened porch and my grandmother set out lunches of borscht and latkes, coleslaw and strudel. And it didn't seem possible to do one of these things without people doing the other things nearby in silence. I waited to hear my father telling the writers to listen for, what to watch out for, when to go back in time and to come forward. "Mainly," he said, "watch people and you'll see your stories."

Now we're in my dressing room. "I love the way I've put all my clothes out on steel racks, like it's backstage."

He says, "It's because the house was built before closets were invented, and," he does this eyebrow lift, "I have more fun buying you clothes than closets."

"You don't have to tell me if you don't want, but do you have money?" I pull out a smoky gray violet outfit with wide-legged slacks and a radical jacket with a bias-cut asymmetric hem.

"That's a new young designer, John Galliano," he says. "And no, no I don't."

This is a subject I feel we don't like to talk about. "But we live like we do." I pull on a red striped skirt that goes with a red tweed jacket.

"It's a Valentino," the Englishman says, "I bought it for you in Venice."

"I loved Venice," I reply. The suit is out now. You don't have to know that to feel it. But telling him that is impossible.

"You don't remember Venice," he's telling me.

"But I can remember I loved it." I look at him. "We can go back."

"Not so easily," he says.

"You'll see," I say, "we're partners. I'm going to work again. My mind just needs some pushups. You'll see."

It's around here I started the Red Valentino System.

Judith and I are eating lunch. She wonders, "How can you write a book when you can't remember what you're writing?"

"I'll remember what I'm writing—I just don't remember I've already written it or where it's going."

"So don't worry about that. Just write it in scenes as they come up" Judith suggests, as if it's a film.

"Some mornings it already feels as if I'm going through a lot full of spare parts, and I'm saying 'make a car out of this.' And I'm no mechanic."

"So, maybe you'll make an interesting car. Or do you want to whine?"

So I write down scenes and as they go, I remember another piece, another part of my life that works here. Maybe I'll put it in the spare parts lot and wander through at the end of a day, seeing if there's something I can patch it onto.

Now Judith's looking over the pages. "You've got a lot here," she says, "but you must have liked that red Valentino suit because it keeps coming up." So I put a red Post-it note at every place I talk about the suit. Then I check which one I like best. With different-colored tags, I do the same system with almost everything.

"Talent," my mother said, "is the only thing you'll ever need."

"Talent is fine. Then you have to show up," my godfather, the screenwriter Lennie Spigelgass, said.

I'd come to him for money. He told me, "Put on the pearls I gave you for your sixteenth birthday, buy a black dress at Macy's, and go and get a job at Saks selling dresses." I did. That taught me fast that writing wasn't so hard. Talent, for Lennie, for Gore Vidal and all the writers he gathered in his house on Sunset Plaza every Sunday afternoon, was the thing that always came first. They'd envy each other, tease and fight, but their lives were also about their work.

I think of my brain like a twisted ankle. It will get stronger. I will exercise it carefully.

10

I open the invitation to Ian's son's bar mitzvah and begin to worry about what to wear. This is built in; worrying about what to wear comes before writing, before worrying what to make for dinner.

"I haven't been to a bar mitzvah, maybe ever," I tell Stuart, who is Ian's business mentor.

"Jeb would have been bar mitzvahed. Ask Joy what to wear."

"She'll ask me why I'm worried. Then tell me, just be comfortable."

"So. Is that so terrible?"

"We get an invitation to a bar mitzvah and you're sounding Jewish. Don't do that," I warn him.

One of my friends says she cannot have sex with Jewish men. They are both too familiar and too couched in the emotional vernacular of their own superiority—and of our superiority. We are like their mothers, and they are like our fathers.

"I think I'll need a hat. Navy blue. And high heels. Can I walk in high heels?"

"No," he says. "Didn't Jeremy have a bar mitzvah?"

I am very still. I must know this. Of course he would have. "Well, for God's sake," I snap, "you were there. You tell me."

"No," he says, "I wasn't there. Jeremy was twenty-one when you and I met."

Twenty-one.

At nineteen I married, left home, and began to learn about life.

My mother sent me long, unraveling letters from New York when they moved there just after Jeremy was born. I hated that my parents had left. She was telling me how much she missed home, but I couldn't hear her. I was too busy missing home myself.

Jeremy once said, "Mom, a real family stays somewhere."

I think of families, places, and memory.

When I visited Johanna's first apartment in Norwalk, I remembered my mother visiting my first place, a Quonset hut in Coronado on the Navy base. She cried. I fed my parents lasagna, which my mother-in-law taught me to cook. My parents had never seen lasagna. This was gangster food, and not Jewish gangsters.

One of the first things you do when you marry is to irritate your own parents with details of the new family's life to show you have moved out. I do the same thing now, but to my children and I get at once jealous and grateful when Johanna talks about her husband's immense, supportive family.

There is no precedent for dealing with the new lives of parents. No protocol for accepting the ways of the new family.

Jeremy didn't really talk to me the last time I saw him.

Stuart shows me pictures from the visit and I remember moments. I remember Jeremy pulled a piece of steak he'd barbecued off my plate and finished it. He cannot hate me that much if he bites where I have bitten.

I watched Jeremy listening to his phone. He's a careful listener. Talking and listening is what he does for work, so no wonder he doesn't call. That is how I choose to see it.

I go with Judith sometimes to pick Noah up after Hebrew school. This makes no amends for what I have guessed I did, which is spend money on Mustangs, not on my children's religious education. Joanne will know and will have a way to fix how I feel. Or do I have to find my own trail here?

This bar mitzvah is in Stanmore at the end of the Jubilee Underground, the long north train out of London. We're talking here about a subway to Rye. I think that's what we're talking about. I don't remember the exact twin city.

Stuart has an easier time with his concept of God—which isn't saying I want to give up mine. I'm not sure I can. It's rooted in there, maybe like the eucalyptus tree in front of our house here. It's looking fairly dead, but I'm not moving it just in case.

"Your fight with who God is," Stuart says when I start this discussion, "may be more about your father. It all starts there, but I'm too tired to go over this again."

Why are women never too tired "to go over it again"? This is not a chauvinist remark. It is the truth.

For the first months, long before writing became a consideration, I practiced reading with Agatha Christie. One book a week. Then I'd test myself. Finally, it was two a day. I am a detective. I see myself slinking along the walls of each day, holding up the magnifying glass. Catch the details and the day will stay with me longer.

Just when I think my memory is all there, I hit a blank patch—always, always, always around the kids. Don't tell me emotion doesn't short-circuit memory.

At the temple, the men are wrapped in white and blue shawls, some wear derby hats, some yarmulkes. Ian's yarmulke is navy velvet with silver, his son's is garnet. Stuart sits stiff and upright with the white yarmulke he's been given perched on the back of his head. The men fling shawls over their shoulders with flair like women with those cashmere wraps Valerie Wade is crazy about. Some drape them over their heads.

I'm sitting with the women in the balcony. Ian's wife, Mira, has her long dark hair swept up and wreathed in gorgeous silk roses. We are leaning over, watching the men freely praying, chanting, and wandering. It's an individual process; some face the wall to mourn, some turn around, some in and out, wandering. It's free spirit. No wonder we drive everyone crazy.

"*Zahar*"—I am looking through the thick prayer book in the rack in front of us. Is this Hebrew for memory? The book says you don't just remember. You ask yourself, "For what?" Remember to what avail? Remembering creates; remembering has action; you must transform the past into a contemporary imperative. And remembrance, here, involves a response to right a wrong. I remember in order to desist, to stop doing something. I shall hold our remembrance. I must remember the covenant I make.

The Nobel Prize winner Gerald Edelman compares the brain more to a dense rainforest than to a computer. That sounds lively, full of surprises. "An ecological habitat," he said, "that mimics the evolution of life itself."

So my memory may not be so much pulling up an image fixed by a program as rummaging through antique markets with Judith and recreating images, rooms, scenes from long-gone decades. This makes memory more like the act of an artist or storyteller, something to be drawn up. I don't have to remember things or situations so much as to be able to do the act, to find the creative trails.

In my mother's painting "Hall of Mirrors," you see her reflection down through several layers in a three-sided mirror, like in a dressing room, only she's set it just so you can't see her face. Catching most memories is something like that. There's an image just out of sight. Petits mals feel like that, like tumbling down through a mirrored tunnel of reflections of past events.

I saw my parents' marriage like that as I grew. I never saw it face to face. (Do we try not to see our parents' marriages? Is that maybe how it's meant to go?)

I watch Ian's wife looking at her son. "Maybe," the woman next to me whispers to her friend, "the reason we keep out-of-date husbands is so we can let go of our sons."

My first mother-in-law knew right away her son's marriage to me wouldn't work. "Do you know anything about being a wife?" She knew. She was married five times. She was between the third and the fourth when we met.

She draped the layers of pasta for her lasagna like a stripper. "Can you do this as well for him as I can?" She sat across from him at the table, her eyes filled with tears. The two families dueled for us. Would my father's projection room outplay my husband's mother's exotic kitchen? The best and the worst thing that happened for our marriage, so that it lasted long enough for us to have our children, was that he was sent off to Korea with his ship.

I wrote around ninety letters a month to my overseas husband, and he wrote me back. His distance was also his particular appeal. When he came home from the service, he was back with his mom, riveted by her forceful, hearty seduction. She couldn't know it, couldn't see it. I couldn't begin to match up to her commanding attachment, her jokes, or her lasagna.

I don't remember eating anything that my grandmother hadn't taught my parents' cooks how to make. Each cook would teach the recipes to the next before she'd go, watching, waiting, until they were just right. The main thing that had to be just right was continuity. No revisions.

MONDAY
Roasted chicken with paprika and egg noodles (not to be confused with pasta)

TUESDAY
Veal cutlets armored in bread crumbs

WEDNESDAY
Lamb chops, always with canned petit pois, stylish in L.A. because they sounded French

THURSDAY[*]
Beef stew and salad (a wedge of Iceberg with Thousand Island dressing)

FRIDAY
Salmon croquettes (same bread crumbs as veal) or filet of sole (Fish every Friday in case the guests were Catholic).

SATURDAY
Spaghetti or hamburgers (our parents were often out),
Or pot roast with latkes, which no one but Grandma ever got right. Which is how she wanted it.

SUNDAY
Deli

[*]If our parents were not having a Saturday party, then it would be pot roast on Thursday. You had to have it once a week, like you had to have deli on Sunday.

"It's your mother or me. Her or me." I remember my mother standing in their bathroom, screaming at my father. "This is my house, my life." She hadn't realized that my grandmother and I had come back upstairs, slowly, after my grandmother had finished helping Ethel make the latkes. I'd climbed into my grandmother's bed and wrapped myself around close to her body in the ways you couldn't with parents. "Careful of my hair," my mother would say. "Don't kiss me like that," my father said when I kissed his smile, trying to make a perfect kiss in one long take, the way Lana could do with Gable. My father pushed me away sharply, and now I'm not sure what I felt then, although I remember my father telling William Wellman that Lana Turner told him Gable had the best staying power of anyone in Hollywood. "And she knows," Wellman said. Staying power. I thought it meant a guy would never leave you no matter what, which didn't make sense with Gable, who kept changing wives with just about every movie after he lost Carole Lombard.

During the sixties, when I remembered everything with the anger of my fierce, alerted sexuality, I remembered that I felt rebuffed, deeply wounded. Now, with more distance and the deeper perception of time and satisfaction, I think my father was startled and aware enough to see my competitive games with my mother and to catch his own reaction. So he was short with me.

Daughters and fathers, sons and mothers; we captivate each other. We are our own other.

I could drive like my father and cook like his mother. These, my mother thought, were not class things to do. Why be seen driving when you could have a chauffeur? Why cook or take care of children? Some women, she'd tell me later in our long night talks, also think sex is not a class thing to do and they lose their men to women who know sex has nothing to do with class. You don't have to know anything else to have a man you love. She taught me early, I think, to keep me from attaching totally to my father.

In 1968 I moved to New York. Work for me was getting the gig so I could tell my father, so he'd notice me more.

I had no money. My second husband had gone there to find better poker games. I'd been fired from my job and had not paid rent for about six months. Writers I knew all said I'd have an easier time getting work in New York.

I had watched my father with his nephews and with my brother. I had learned everything I needed to know about the most important things from my parents: romantic love, how to use my talents, and how to work a room. This is why I probably never set much store by schools; neither did my parents.

I liked the bits of fame I could grab. Life was superficial, extravagant, edgy. Even after I learned to listen to stable women with strong marriages and strong relationships with their children, I used my "adventures" as party entertainment.

I didn't talk to my children about where we were going or why. I held them close until they showed interest in the outside world, and then I brought that world in on them in my version of their terms. They were my right and left arms, sunny companions. I raised them in the sentimental Marxist style of hippies—a platform I fell off every time I got my hands on money and near a Mustang. Then we became the Three Musketeers, triumphant with our cars.

I rationalized that I had no use for the closed-in temples and the dusty attitudes; God was in the trees, the curves of the mountains, and you could get close to God best driving on those mountain roads in your car.

With the loss of memory goes the loss of preconception, the loss of references. Then, just when I think the door is locked to some area of my memory, I find a key and discover another rocky old road to climb. If I'd seen all of it at once, I'm not sure I'd have made it. To reach some memories you just have to stretch harder, practice longer, then try again. A lot like the laps in that pool. (And I remember the pool even as I think this through.)

The rabbi at my confirmation greeted me with his palms around my forehead as I came past for my blessing—telling me, "I would have given you the confirmation cup if you'd been a good student because of everything your father has done for the synagogue, but you haven't deserved it."

I hated this formal, self-righteous leader handing out his blessings like traffic tickets. I'd been charmed in the beginning, but that day I felt trapped, locked into a framework, my idea of God all twisted around with how much you'd give for the new benches or whether I'd be pretty enough to marry a man who could buy a whole table to the dinner for the man who donated the new wing.

At the bar mitzvah lunch, we are sitting at a table with eight men Stuart knows and only one other woman. "It would be a minyan," she says, "if you and I weren't here."

"Do you have any children?" Sam, the man on my right, asks me.

"I have a daughter, Johanna, and a son, Jeremy."

"That must have been some bar mitzvah, out in L.A."

I'm sidestepping this. "Where was yours?"

"I didn't have one. My father died when I was a kid, which made me mad. I didn't go to his grave until I was around twenty. He died at thirty-four. I had trouble finding the grave. It was covered with shrubs, vines, and dirt. I found a service to keep the grave clean, but I never went back. I think of my father around some of his birthdays. As time goes on, he has a different role in my memory, do you know what I mean? When I was thirty-four, I saw him as he was when he died. I felt as if I was losing an older brother. Later, I felt like I lost a younger brother. It's harder to stay angry with him now that I'm forty, really old enough to be the father of the young man my father was when he died, if you see what I mean. How old is your son?"

"He's forty now." I'm trying to figure out how old he was when I was forty. This is not memory. This is not being able to do subtraction in my head. Using my fingers, I've figured out he was eighteen when I was forty; five years earlier he was thirteen. That was 1971. Rough years.

There's a message from Judith when we come home. She's got a late-night shoot. Can Noah come over? She reminds me when they arrive that he has homework to do.

"Sure," I say, "right away."

"I mean it," she says.

After Noah paints a giant charcoal robot on his stretch of the wall along the staircase, I say, "Time for homework." I never remember saying that to my own children.

"It's always time for homework," he says.

"The longer you put it off, the more time there is."

"That doesn't make sense."

"I know, but that's what you're here for."

Noah needs to see where letters go. I need to see where my chapters go. Not to mention the characters and their situations, which become more baroque each week as I forget what they have been doing.

We're lying on the floor of my workroom. "I need some more marshmallows," he says.

"And I need another Porsche," I say.

"What about one marshmallow?"

"That's it. But only after you spell 'economic.'"

"What does it mean?"

"It's about money. An economic issue is whether you have the money to get the Porsche."

He spells Porsche in an instant. But the Porsche has a picture that makes the spelling stick in his mind. Even in mine. We make a sentence out of "economic": Elephants Come Over Now On Mondays In Coaches. "So, if you're in a place where spelling *economic* becomes a big deal, you can pull out that image like a kind of video of elephants waving their trunks, leaning out of a huge coach as it comes up your drive."

"One of them might toss the Monday paper on the doorstep," Noah says. "The way I spell economic is I use a laptop with a spell-check."

I don't have space in my neurological attic, my changing room, for this, but it is there, trying on image outfits. The elephants are Republican. They are wearing narrow ties, fanning themselves with folded-up copies of *The Wall Street Journal*. Where did I find that?

Noah's ahead of me in some areas. He can tell you where he lives without even thinking about it. But I spell better.

"However," I tell Noah, "I'm working out a system for spelling, which isn't my idea I'm sure, but I think you're going to like it."

"Could I have one marshmallow?" He goes into the dining room to play with my father's antique banks, while I try to find where I've hidden the marshmallows.

Noah's taken down the bank that actually looks like a bank, with a teller there to take your change when you open the door. I pull a chair over to the window. "Hand me the tape," I say. I gather the fabric falling down from the valance into my hand and put it back up with lassos of double-sided tape.

It's four in the morning. I'm sitting on the edge of the sofa in my work-room. No, you can't call Jeremy and go into tough subjects. Such as bar mitzvahs.

I can't find any characters around, no one I know. I set up a scene. They wander in and stand there in Dacron suits, hands limp at their sides, people uninvited to a party they don't want to go to.

I ball up pages. Throw them across the room. I can call that man who copies instruments, objects, or is it genes? He's from Fulbright—or Oxford? Look it up. What's a passing connection becomes a research project. Yes, here he is—John Halloran. He inserts real things into a computer and reproduces them.

I call Jeremy.

"It's four in the morning there, Mom. What are you doing up?"

"I can't write. I think the doctors knew it wouldn't work."

"Don't you remember what Grandfather told his writers? I think I've told you." He probably told me and thinks I'm pretending I've forgotten so I get a chance to call him. That's fine. "I know I've told my writers and it works." Like Lynn Nesbit's, Jeremy's voice changes when he talks about writing. The tone mellows and the beat slows way down. "Just write your characters a letter, Mom, and they will answer." Pause. I can tell he's look-ing at something else. "Is that all?"

"Yes, that's great."

And he's gone, and I forgot what I really wanted to say.

"You're not sleeping," Shaw points out. He's come into my workroom around four this morning.

"That's true. I've tried Jeremy's advice. I have written to two of my characters. I have heard nothing."

"They did not have airmail," he says, "it will take time."

"Don't be funny," I tell him, "it doesn't work for you. You are sexy, interesting. Not funny. I would never sleep with a funny man."

I am afraid, I explain to him, that I've been away from the characters too long. "When you make someone up, their presence can just go, like smoke."

He reminds me I never really make anyone up. "Your role models this time are L. B. Mayer and the original moviemakers."

"In a way," I tell Stuart, "it's coming clear. My father used to sit around a table like this with his writers, Millard Kaufman, Allen Rivkin or Norman Corwin, talking characters alive. I'd like to write about a female mogul, a Doris Vidor or Frances Goldwyn, who inherits a studio."

Stuart says, "So just imagine Judith on a ranch in L.A. in the early part of this century.

"Listen," he adds, "I came in to tell you that it occurred to me, if I were a writer and I'd lost my memory, one thing I'd do is reread what I'd written."

"Did this occur to you while you were sleeping? I like being part of your unconscious."

"It was simpler. I woke up and you weren't there, so I knew you were worried about the book. Look at the pieces you've written for the *Times* in the last few months."

Silence.

". . . on teas in London . . . on the new-style country house hotels . . . "

This kind of short past time collapses like an old balloon and flits off.

"You know," he says, "your writing is best the closer you stay to reality. Always has been."

He doesn't need to add any more to this. But he knows I would prefer to spend a lot of time discussing what he means and how to go about it, so he sits down in my Aunt Lillian's old rocking chair across the workroom from me. As he leans back in the chair, one of the lights in the wobbly old sconces by the fireplace goes out. "I thought we had that fixed," he says.

"That might have been the other one." The light snaps on again. "It's best not to notice."

"Fine," he says, which it is with him. He's not drawn to fixing things. He has this toolbox in his changing room that serves largely as a mnemonic aid for both of us: me for putting memory test objects into; Stuart for reminding me how distracted he'd be if we did live in the old house in Connecticut, where it is the custom to fix what is broken.

Sitting here now, Stuart crosses his ankles exactly the way he did when he was four. His mother gave me a picture of him then. As we grow older, we look more like ourselves as children, or maybe we just try to recall all the possible ways we can be appealing. Little gestures and expressions that got surefire reactions. He is not conscious of these things; I probably am when I do them.

"I guess I ask you about every page ten times," I say. "You're very gracious about not telling me when I repeat myself. Or maybe to avoid going nuts, you just don't pay attention. Is that true?"

"That's not an argument I can win."

"I think I see what you mean." I've half-forgotten the question.

But I have learned when I repeat myself on subjects he's bored with.

It's not exactly that I remember the repeats. I recognize disinterest, like animals being trained to go through a maze by electric shock. In this case, the door of interest snaps shut, so I turn to another subject.

"But you're always great about my work."

"This is true, even though you're complimenting me now to keep me here with you."

"I could go over and stroke you, but then I'd go back to bed with you, and I want to work—have to work."

"I know that," he says.

"When did you figure that out—right away?"

"No, or I didn't exactly. When I first met you, you took me to your therapist, to Gloria . . ."

"Gloria Friedman." I haven't called her, I remind myself. I can't imagine why I wouldn't have talked to her right away. I'm interested now, too, that I can think about that and listen to Stuart at the same time. Have I even forgotten how thinking normally works, that it does fold over as you talk?

"Gloria was very wise," he's saying, "she said everything would work out fine so long as I understood that the book you are working on always comes first."

"Did I remember right away that I am a writer?"

"I told you right away. I knew it was what Gloria would have said to you."

"Would have?"

"She died a few years ago," he says.

I look around the workroom. Posters of my father's plays, a drawing of Ethel Barrymore, and a picture of an ancient forest. Did I put all this here in case the memory went?

"I remember Gloria. Mostly I remember she was always there." I think about it. She'd be there, shifting in her chair with pleasure when she had an idea that would help. "You're always here, too," I tell him quickly. Am I here for him? Do I register what he does with a day? Does he know how little I remember of our years together? How many, how few those may be?

"You mainly have to remember work never just jumps onto the page. You have to trust it will come," he says. "It takes a long time, but you've known that all your life." He gets up now, comes over and kisses my forehead. "Your characters are answering you now. I'll see you later."

That the work takes time I learned from watching my grandmother with her knitting. Even before the war, she made dozens—more than dozens—of caps, sweaters, booties, and gloves for refugees' children. I knew how much time her work took and it always got done. Her hands were always filled with wool, or food—challah, knaidlech, matzo balls—twisting, smoothing, forming it. When she lived with us during the war before my father got her a flat of her own in Westwood, she taught my mother to crochet and knit. They'd sit listening to war bulletins; my grandmother would hold up her hands and my mother would take a new skein of wool and wind it around into a tight ball.

My mother's hands were also always moving, across piano keys, sketching, dappling, sweeping a brush over a canvas. "I'll be okay if I don't try to keep my hands still, doing nothing," my mother said, "the lunching ladies do that. They can't break their nails." My mother spent all day in wedgie shoes and blue denim aircraft workers' overalls the studio fitted for her. She'd wrap them over with denim skirts later on; my father hated the actors she painted seeing the outline of her body.

I look up at the top shelf across from me. I have written five books: *With a Cast of Thousands; Thanks for the Rubies, Now Please Pass the Moon; Bed/Time/Story; Perdido;* and *Dr. Rocksinger and the Age of Longing.* One by one I take them down. They all have "I" voices. So, in this way, I should be no stranger.

But each "I" voice adjusts, more or less, to the slang of her time.

The first book was a cheeky, naive little memoir about life in fifties Hollywood. How blind I was. And it wasn't that you just didn't write about those things; you didn't know. I see a smothered kid here in *With a Cast of Thousands,* which was written in 1963. It was right before I burst out. I wrote about the budget my father showed me for his first years, which included entries for gum, cigarettes and writing paper, the R. H. Brothers Tropical Furniture, and the giant Capehart radio phonograph that you'd sit and watch, like Ishmael, the gorilla.

I say in that early book, "My mother was repelled by infidelity, homo-sexuality and rudeness." Where did I get that? Her closest friends were gay, rude, and played around. I wrote the book for my parents the way they'd want it to be.

It says, "I was a pig about attention—even bad attention was more welcome than none." My mother, I'm saying, knew the threat of boys talking about me wouldn't have kept me a virgin—only what my father thought of me would. I've always had to have an icon, someone my actions play to.

I have my grandmother in every book, from the Witch Neva Sheba in my second book, a fictional autobiography of Jacqueline Onassis (edited by her godfather, whom I modeled on Gore Vidal, a friend of my own godfather), to the Rocksinger with the silver hairbrush in my fifth book (can you have a lover and a grandmother all at once?). Grandma was in *Perdido*. And in every book I talk about driving and hating to walk; right here, in *Bed/Time/Story*, the husband of the time tells me I don't have to take the kids five blocks to school in a taxi. We can walk.

Suddenly I remember "pig" from writing for the *L.A. Free Press* in the sixties, calling cops "pigs." Or was it thinking of my grandmother and "trayf"? I should simply write about my grandmother and her life in Hollywood with my father and his young friends.

I start to run to tell Stuart I've got the story, then I hear him downstairs, alone, playing the piano. Do not smother him. There has to be a time where we can each see our own days. I don't want to think of that.

It may be the next evening that we're lying on our bed watching *A Farewell to Arms* on television. It's a period that moves me, maybe because when I try to remember my parents and who they really were, it helps to go back to their time and imagine the impact World War I had on their lives.

"I think I'll keep this book closer to my parents' lives, research and talk to people they knew, and then make it how I see it. I can't go back to them, so I'm not locked in by the need to get it right. It's a closed story. I know the ending."

"It might be a relatively more urgent idea to look at your own life," Stuart says, "to see how the shifts and changes hit you and your children—and changed all of you."

"That's also a closed story," I say too quickly, because I fear it is. "And I wrote that book— about the woman in love with the rock singer not much older than her kids."

"But you'd tell a very different story now. Your own story is changing as your needs of it change. The emphasis is different. You're not lying, not even forgetting; you're accommodating their own changes."

"Are you being tough on me?"

My writing won't fix anything with my children. What they'd want is for me to understand how they feel. And the years that affected them are the years I can't catch.

This scene from *A Farewell to Arms* is set in a wooden mountain house near water. "It's probably the Adirondacks. I lived there once," I remind Stuart, "the kids loved it."

They came back to me there after living with their father. I made some money, which I spent in about an hour and a half; as if money would fix it. Band-Aids on deep knife wounds.

"It wasn't the Adirondacks," Shaw says, "you lived in Connecticut in a place called Stonybrook, near Ring and Frances Lardner, with ferns and green slopes, pine and evergreens. You liked living by another writer and you said it reminded you of a canyon in L.A. We lived there for the first three years we were married."

"So, was it like this?"

"A little bit."

"Did my father visit us in Connecticut?"

"Yes. You bought the house because you thought he'd love it. You were also showing what you could do. Those were not easy times for your father and mother. They came for Thanksgiving the year before we married, the year before he died. But I wasn't living there yet."

"Why?"

"We didn't want to live together until we were married."

"That was smart—to get the children used to the idea."

"Not exactly. It was to get us used to the idea. I was used to complete independence and you were used to having total control."

As we're having this light dialogue, I am walking through the Stonybrook house, placing my father in different chairs. Light from the skylight falls on him as he comes in and sets down the bag he carries with maps and stuff on the round breakfast table. I've put a Western scarf over the table, which he'll like.

I see my father sitting in the screened porch in that Stonybrook house telling the children stories about when he was young, the way he did with us. But there never was enough. There never is when work takes its cut of your parent's day.

This is some kind of triumph. I am here in London talking to Stuart—and I am seeing my father in Connecticut. I could not do that a few weeks ago, could not layer thoughts of the past with conversation or an awareness of the present.

A few weeks ago? Well, whenever it was that I couldn't do that.

12

I meet Judith for lunch at Villandry, a little restaurant full of rough tables, big helpings of Provençal food and the only perfect raisin challah in London. People in film and the rag trade and people from Hampstead have jammed in here since it opened up the street from where we live.

Jean Charles, the owner, tells us they are moving. "You are having our last meal here."

"I've seen the new place," Judith sets her phone on the table, "it will be very good." Then adds, "I'm moving, too."

"That's sudden," I say.

"It will make life simpler," she explains. We split another piece of challah. "It's on Cesario's street."

"Even simpler if it was in his house."

"I already told you. He's married."

"Then his wife will be pleased when she gets the change of address card."

"I thought so," she says. Before I can tell her I don't think this is a brilliant idea, she switches to a subject that will distract me. "So, are you okay?" she asks.

"Imagine riding horses who go all over the track," I say. "That's what it feels like inside my head."

It's easy to describe. I just don't like living with it.

"Hard for either of us to imagine riding horses," she says. "Let's say a guy on speed."

"Right," I say, "but I can make myself come up out of one of the attacks. I say 'stop it,' and it does stop, the way I'd not throw up, or stop coming," I tell her.

"We all know how to do that," she says.

Starting, I don't say, is the usual trouble.

The phone rings. She hopes it will be Cesario. It isn't. She's quick on the phone. "That's my friend Geraldine," Judith says. "She's trying to get her life together and manage a rock group. Actually, she wants to be a writer. Maybe you can give her some advice."

"Take up skiing."

"Exactly what I thought you'd say," Judith says. "Maybe I should call Cesario."

As I listen to Judith, I am beginning to remember how very much of my life I have spent on the subjects of sex and waiting for the phone call I wanted.

In every stage of life, there is always someone who has not called. Call it a growth requirement. In the First Act, it is parents I was waiting to hear from. What do they think of the drawing I slipped under their bedroom door? Do they like the card I made? When are they coming back from their vacation? (A holiday was when I went with them.) Is that Lucy, my father's secretary, to say he's on his way home?

Second Act. The friend I loved the most; the guy I loved the most, longed for the most, will not call. I didn't go out in case he calls, or she calls. There was a time, in the dark ages, when there was only one phone. My sister and brother were designed to be on that phone, to keep the call I wanted from Caroline Veiller or Paul Sperry from coming through.

This was also a time when I couldn't go out because the phone could not go with me. Not that this helps. No matter how many phones I have, I am on all of them when the call I want is trying to get through.

And when it does, the sharp, funny things I mean to say—all offhand, throwaway lines that show I couldn't care less—are gone. Breathless? Never. I used to get asthma when Paul called. I said it was because a cat had just walked by.

In this Third Act, most of us have more or less worked that out. Or given up. Which is a way of working it out. This does not mean I am not waiting for the phone to ring.

It is Sunday. By four o'clock I start waiting for my kids to call.

Ten, fifteen years ago, I hated to pick up the phone on Sunday. "If it's my mother, I'm not here," I'd say.

She'd know by the way the phone rang that I was there. The way Stuart's mother knows when we don't pick up the phone on Sunday mornings.

I look at the picture Stuart tells me is Jeremy. This is a picture of a good-looking young man with a drill sergeant's haircut. "Tell me he is not in the Marines." He has the stark presence of a guy you'd see in an industry ad. "Industry" means Hollywood. This is on the hard disk.

"He doesn't like talking on the phone," Stuart has explained. "For an agent it is not a recreational activity."

I decide Jeremy is a combination of the characters on *Friends*. It takes more time than you'd imagine for me to get clear that my daughter, Johanna, is not like the Phoebe character. Johanna, I learn by doing phone brunch with her, is also witty, but earthy and stable. She outgrew whimsy by age ten. Phoebe is also not like Jeremy's Phoebe, whose mom, Romi, I decide, reminds me of Jennifer Aniston.

Don't try logic on it, I say to myself when Stuart switches the channel. Just let it be and you won't pine for whoever they really are. I tell myself to assume that my kids have found worlds as well that are even more reasonably real—could I dare to hope, these days, even actually interactive? And what do I mean by interactive?

After living with no memory, writing postcards I didn't send, and rereading my parents' papers, I decide I might write again the same way I learned to do something else I didn't think I could ever do—to live without speed and drinking. I'll find other people who are trying to do the same thing—to write.

"They could come every Sunday, maybe for lunch," I tell Stuart.

"Maybe," he says, but he isn't listening now.

I want the writers' gathering to have something like the feel of the Friday afternoons when John Lahr held his teas at his house on Primrose Hill, before he began to review for *The New Yorker*. Here people like Betsy Blair, Karel Reisz, and Al Alvarez all gathered around the kitchen table.

Betsy and I threw our arms around each other when I first saw her. She smelled of smoke, of the forties, of my mother. She was married to Gene Kelly then, and I remembered when she posed for my mother in ballet tights.

I already knew I had met my first writer as I spoke to Angie Montague at one of Laurie's exhibits. Angie said she was a poet, but was interested in writing a novel about a painter. A tall blonde, she was trembling with the particular fear it takes a lot of talent to get. You have to know how hard it is to be good.

And she understood right away about the memory.

"I may not be able to remember what I read," I warned her, "and I won't remember you . . ."

". . . or what I'm here for . . . ," she smiles, bashful and wise all at once, "when you leave the room and come back in."

"It can be that bad, yes." She arranges her words, speaking in short, careful lines.

I came home one Sunday afternoon after lunch to find Angie sitting in the large rose velvet chair in the black and white marble lobby of our house. Laurie had let her in. I had forgotten this was the first Sunday we had agreed to meet. She had her arms around a manuscript she didn't really want anyone to read.

"I don't want to read mine, either," I said, "so maybe we'll just talk and see where it goes."

We decided to listen to each other read and make comments as we went along, catching when it doesn't sound just right. Then Angie suggested that during the week, she could help me organize the notes for the book about my parents, which Stuart tells me I'd started years before. This is why Boston University sent me the cartons of papers, letters, pictures, and manuscript pages that are in my workroom.

Angie manages to be totally supportive and needy at the same time. After about a week, she could figure out where I'd be if I wasn't at home when she arrived. I don't know I have a pattern, but everyone else does and can usually find me when I've forgotten where I'm supposed to be.

She helps me to set up a chart of scenes leading to the Happy Ending I am sure my parents' story will have. I will pick up an Oscar for my screenplay of this book about my parents. My father will take off his glasses and, with his head to one side, wipe his eyes the way he does. This is not true. My father told me to stay out of his territory.

It is only a few weeks before there are three of us here for Sunday lunch, talking about what we are not writing and reading what little we have. I met Geraldine at a screening at the British Film Academy.

"If nothing else," she says in her gentle Scottish brogue, "this will be a distraction from making unsatisfying phone calls." As the road manager of a rock group, the novel she is writing makes clear that the phone calls might be to the rock star who is married to her lover.

Geraldine has strong shoulders and long legs. I can see all too fast why she appeals to everyone.

"The book's not what I want it to be, but I'm just writing and that's how it is on the page," Geraldine says. "There's something wrong, but I don't know what it is yet."

"It's probably an easier story to write than it is to live," Angie says.

"Especially," Geraldine says, with a wry charm the brogue adds to everything, "since I know it's not going to have an ending I'm going to like."

After three or four Sundays (do I remember?), Stuart asks, "Are the girls coming today?"

"It's not girls," I say, "Mark's coming, too. And so is Judith."

"I don't think that's a good idea," he says. Stuart sees himself as Judith's mentor, and he's probably right.

The only really good idea, for Stuart, is to be alone with me on Sunday. Quiet. But he also understands that I have to collect attachments to hold in the wings. God forbid I should be alone and have no one here to say, if not "I love you," at least "This is who you are."

"I'll take a walk and see you later," he says.

A walk. Of course.

Angie and Geraldine both understand that when I forget things they tell me, it isn't because I'm not interested. As I set out lunch, I let them read a study my sister, Joy, has sent me from the *American Journal of Psychology* on "The Effect of Activity upon Learning and Retention in Cockroaches." Joy has attached a note, "Thought this might be helpful."

"Cockroaches don't like walking," I say, bringing in the bread, "it disturbs their memory."

"That's why there are no cockroaches in London; they'd have to walk." Mark glances at the study. "This is from 1947, before anyone heard of fitness!"

While Geraldine is showing Angie more notes she's written for her book, my phone rings. "Don't," I say to myself, "don't pick up until the third ring." I can't let Jeremy know I'm sitting here, waiting for the phone to ring. Maybe I should let him think I'm not home, that I do have things to do. "No. Pick up the phone."

Pause.

It's for Stuart. I go back to the kitchen. I have turned on the wrong burner, so the chicken broth for the risotto is still cold.

An attack starts. I hold it out there to look at it, the way you'd step back from a painting. Tell it it's just fear. Cut. Rewind.

"Do you want a glass of water?" Geraldine has walked in. I'm sitting at the kitchen table.

I can't answer. I see flickering images of Anatole dancing with Bliss, dappled Connecticut trees. I want to say a name. I can't find the name. Trigger. Roy Rogers's horse. His wife's name. Taste melted cheese sandwiches. Dale Evans. That was his wife's name. Now I see old trees bent by heavy Connecticut winters. Heads wrapped in gauze, the smell of Shalimar, black cords, and steel clamps.

"Isn't there something you can take?" Geraldine puts her arm around my shoulder.

"This Dr. Earl suggested I up what I'm on to four hundred. I would be completely unconscious and stoned, a condition I've worked rather hard to avoid."

"I can sympathize there," she says.

"I thought you might," I say. You can't be the manager of a rock group and not have a deep understanding of drugs.

"Then don't do it." Geraldine says. "You have to think about it as one more risk."

"But there are some risks not worth taking, you know." I can't forget my role here.

"Sure," she says, "but if you know the answer, if there is no danger, it hardly qualifies."

"Could we talk about danger and you?" I ask.

"Maybe another time. You can't believe how exciting it is to know I'm really going to learn how to make my book right."

"We're learning together." I'm careful.

I can only guess at what I teach, which is that even if I did remember how to write, it isn't easy. I've found enough of my own notes to tell me that. I explain that, but Geraldine's glowing, her light green eyes looking sideways.

"I'm already thinking about new ways to remember it, to put it down, that make more sense. Maybe," Geraldine takes a bite of salad and carries a bowl to the table, "just the act of being together, saying we're going to write, builds the confidence in all of us. I can feel a kind of spirit here."

"Oh please," Judith has just arrived, "don't do that." She has an armful of spreadsheets for Stuart to look over.

While Geraldine and Judith decide whether they'll tolerate each other, I start the risotto again. Mark, who drives me around with a black cap on when he's blocked and can't write, comes in with Laurie, who sits observing our gestures and expressions. Mark adds some fresh basil to the salad. He's sad, and I can't remember why.

Stuart has returned from his walk. He talks to his mother on the phone and comes in for a bite to eat.

"She's fragile," he says to me. He deals with all these people being here by talking to me as if they aren't, at least until he decides what role he wants to play in this domestic scene. Will it be Hamlet in the corner, or Henry VIII, the burly and vigorous host?

"We're going to have to go visit my mother next week," he continues. "She told me I haven't been back since I left for America. She can't remember the last time she saw me."

"How disconcerting," Mark says, "to have it on all sides."

"You'd wonder if you exist at all," Laurie says, "if every time you turn around, someone has forgotten you."

"Maybe you should be the writer," Geraldine says to Stuart, "it's one way to keep track of what you want to remember."

"He is a writer," Mark says. I wonder if Mark sees Stuart becoming a martyr, dutiful and trapped.

"And a spy," Geraldine adds. "That's how Jill said you came to America."

"Jill told me he came to New York as a poet," Laurie says.

"Thank you, Jill," Stuart says. "Actually, I went to Washington, D.C., as a writer and got involved with the Army-McCarthy trials. When the trials were over, I saw an ad for a job at Procter and Gamble. I thought it was a writing job. In a way, it was. I got the job at headquarters in Cincinnati, Ohio, and with luck, I suppose, and my distinctive accent and journalism experience, I transformed myself into a man in a gray flannel suit. I was rather successful as P and G's first foreign executive in the U.S. business world."

He stops for a moment and says, "If you're writing a memoir, Geraldine, the reader wants to feel the writer reaching for the truth."

I'm stirring the risotto slowly, the rhythm coming back the minute the wooden spoon touches the pan, swirling the rice through the heavy green oil, adding the broth.

"But our perception of experience is so individual," Mark says. "My memory of today is completely different than yours will be."

"That's why you write your memoir and I write mine," Stuart says. "If you're going to be a writer," he says, shifting attention from himself, "it has to come before everything else."

"I have a friend," Geraldine says, "who's an Arabian horse dealer. He loves my stories more than anything—or more than almost anything."

After Judith leaves, offering to drive Angie home to Clapham, Mark leaves and Laurie goes downstairs. Geraldine helps me clean up. She says she thinks Mark's very sad. Stuart says Mark's partner is dying. They told us when they were here last week. I'd forgotten about that—set it aside completely.

Then, after everyone has left, I tell Stuart, "I hate the doctor I saw yesterday. Let's look for another one. How can I be a part of anyone's life if I can't hold onto what they need me to know?"

He's wondering about that, too. "You didn't see Dr. Earl yesterday," he reminds me, "and you're switching doctors like you do hotel rooms."

"And tables at restaurants."

"That," he understands, "is in the genes."

I am looking at the pictures on the kitchen wall. "What if I have a thing in front of my children?" I will not say "fit." "I'll never be able to take them on drives."

He's patient. "The grandchildren, you mean. There are other things to do with kids than drive them around."

"Teach them how to paint on walls, for example," I say.

"You can if you want," he replies.

Jeremy and Johanna will never let me be alone with their children. I can barely keep track of myself. I don't see any way to look at it that I like.

Stuart says he wants to pick up some more milk. We had some delivered this morning—I can see it in the fridge as he's telling me. He needs to get out. I'll leave him alone when he comes back. Maybe he'll write.

I hear him come home. After a couple of hours, I come downstairs. He's listening to Carly Simon and looking at pictures of a woman. "I'll understand if you leave," I say to him.

"Don't be crazy." He puts the pictures back in an envelope.

I reach for them. "Who is she?"

"It's Carla," he says.

I'm blank.

"Carla's dead. I knew her before I met you."

I don't even remember my rivals. "Well, you couldn't have known her after we met, because I'd have killed you."

"Maybe I was rehearsing grief," I hear him say.

"Maybe you were. Welcome to my family—it's catching. I should have told you. This isn't easy for you." I stand behind his chair, my hands on his shoulders.

"I'm afraid of feeling fettered," he says, "of my world closing in."

"That's why you take all these little walks. Every time I turn around, you're gone on another errand."

"Maybe you've forgotten that I always come back," he says.

"But you come back, nervously, like you'll find me on the floor, biting my tongue, pants wet, eyes rolled back in my head. It can't be pretty. But I do remember you saying, like Churchill, not to surrender. I'm not going to," I say to Stuart, "don't be impatient with me, but I'm not upping the pills. I'll take the risk, and the little attacks are okay. It beats asthma."

My life is about my work.

It's four in the morning. I walk into my workroom. The notes Stuart has shown me are lying on the desk. Reading whole sentences is a problem. Pages are almost impossible. I can't hold onto an idea. "Excessive irritability hinders the correct response." I have read that. Somewhere. Or was that something one of the doctors said when I said I'd kill him if he told me I couldn't drive?

I look through my notes for a character to meet. Then I wander through my mind, like a vacant arcade; faces peer around columns, like characters in a hide-and-seek game. I am at Chadwick, at boarding school. Lennie Schreiber is riding. I am on the ladder painting the mural in the dining room. Then I'm going home in Susan Simon's limousine. She's not Susan Simon now. How can I call a friend and ask, "who have you grown up to be?" I'll call Jeremy. I'm clear on him tonight.

But he'll put his hand over the phone, saying to the person in the room, "It's Mom."

I call Jeremy. I hang up. It's four in the morning. "I'd be one of those crazy mothers who call people at weird hours and say I forgot what time it was," I tell Laurie the next morning as she comes up to fax a note to her gallery.

"But that's when you always call Jeremy," she says, "it wouldn't seem crazy to him at all. In any out-of-line way, I mean."

I remember the feel of that night. I know where it goes in my story and where I'm going from here (most of the time). This is a very big deal. I give Dr. Zilkha credit for this.

I liked Dr. Zilkha right away. A tiny, portly, electric man in a suit and bow tie, he comes out to greet his patients. A child has been waiting before us. "I'm just your size, aren't I?" he says to the child to put her at ease.

It's my turn. He tells me, to put this child at ease, that not only skills like typing and drawing will come back, but writing will, too. "The brain likes to exercise." He is refreshing. He says I must sleep more. "Some animals have sharper memories and learn more quickly when they get more sleep."

"That's all I'm doing," I say.

He goes over my records, "Not drugged sleep—you're on the wrong medication. Tegretol will help, and you will write."

"Did you know," I tell Judith, pleased with my memory, "goldfish and dogs like a good night's sleep. Rats and cockroaches work at night."

I am looking out the dining room window, down into the little mews below, empty, silent, big old garage doors and empty hooks and chains for hanging planters. Lilia has shown me where I have hidden the key to the silver chest. "Madame, you put it up here," she pulls out a chair, "so no one, not even I, will know where it is. Up here no one can see it." She pulls a chair up to the bookcase, stands on it on tiptoes, and lifts the key out from behind the dried flowers.

"And it is here," she says, pulling out another drawer, "we have all the bandannas for serviettes." Serviettes? Yes. Napkins.

"Every color." She's probably right. "Most of all red." They're folded neatly, color by color, in a drawer.

"My father keeps his silk handkerchiefs just like this," I tell her, "each color monogrammed and placed in its own stack."

"I show you something," Lilia says, leading me upstairs to the Englishman's changing room. With a magician's flourish, she pulls out the small top drawer in his dresser, "and here, Madame, the very same." And, true, here are silk squares, each one in its place in the spectrum.

I confuse my life with the lives of my parents. This way, they are not dead.

A woman stands with her back to me, looking in a three-way mirror. She is painting a portrait of a child standing on a tall chair, like a bar stool.

My mother was a painter. Equally crucial was her facial paralysis. Half her face was glowing, expressive; half her face was frozen and grim, the eye exposed, unable to close. I knew she was always watching with that eye, even when she slept, and certainly when I slept. So my mother saw everything. She never missed a mirror or a reflection. She always checked to see if, maybe, it had gone away.

I am on a stage in the halfway land between reality and petit mal.

"Are you all right?" Stuart says it again.

I have had a couple of these things already today, but this one is hard to get out of. I hold on tight to his shoulder.

I can't answer. I'm in this glass bell, a film's stuck. Now the voiceover commentator's at the track: "They're off! The ribbon man's tying her down. The hat's too tight. L. B. Mayer's white house is coming into line. Here's his new wife. There's a runaway horse. Sharon Disney's passing a plate to Nora Ephron and, on the right, we're sliding down the wet ivy, spinning down."

Now I'm back.

"Sit down," he says.

"No, I'm fine." I'm clinging to him. "It's just one of these things—like time flips—well, more like . . ." I try to imagine how to say it as it happens again, to catch it and hold it, looking at it like one of my father's crystal paperweights, from every angle. "Imagine," and as I say this, I think I have said it before, which does not stop me. "Imagine your mind about to throw up, on the verge, and again, on the verge, and once more, whoa, a swoop, and it's one of those loop-em turn car rides I'd never go on and I can see why not now." I've had these things for years. I thought they were leftovers from speed, brain damage you learn to live with. Suddenly, defined, they're scary. Suddenly, I understand they can go further.

"I was reading somewhere—today, I think," I say, "that you don't forget skills, like typing. I wonder if cooking's a skill?"

He's being courtly and careful. "I think it's great that you remember what the doctor said. And you don't have to wonder about the cooking—this you have not forgotten."

The slow swirl of the olive oil into the cream face of the cobalt blue Le Creuset pan takes my mind off the rent. The rental agent, Mr. Barclay, has a crush on Laurie, and I know that because she's the only one fastidious enough for Barclay and it keeps him coming back, giving us hints on ways to keep the house. Are we losing the house? Where does that come from? Don't ask. You never mix money with cooking.

I first thought I'd learned to cook from watching my grandmother Belle make lunches of gefilte fish, borscht, latkes, applesauce, and chicken for the young comics home on furlough, like Red Skelton and Jack Benny, who came on Sundays to pick over her Yiddish stories and jokes, the way the kids picked over the cloaks and hats. They longed to be sexy and handsome, and were stuck being funny. My grandmother directed them like a master chef with a brigade of sous-chefs, showing the moves, the eyebrow angles, how to do comedy with top-of-the-chart grace.

Though my grandmother was a storyteller, my mother was a listener. Stars would come to her studio to pose and tell her their troubles. The wide, high window slanted light across the huge room of warm, unfinished wood, where big canvasses hung and slid into racks and two huge easels, like the gallows in *Tale of Two Cities*, rolled across the floor. They'd choose something from her baskets of costumes and props, comfortable, wearing what they wanted, and feeling like intellectuals—with the smell of paint, with classical music playing in the background, the stars would tell her everything.

My mother started out to be a concert pianist, but when her paralysis hit (was it a fall or an illness?—it was never clear), she refused to go onstage.

"Remember," Joy says, "the night we were watching rushes for *High Society*. Dad looked at the close-ups of Grace Kelly and kept saying, 'That face. Look at that perfect face,' and I looked back at mother and she was devastated."

My mother had studied at the Art Students League and worked their formal technique with ease, like a seasoned chef, slipping in the right proportion of linseed oil to turpentine while she was talking, the way I do when I make a salad. I did learn the flip confidence I have with cooking from my mother, even though she never went in the kitchen, claimed she didn't know where it was. But now she's here, perched on my shoulder, with the golden art nouveau roach holder Gloria Swanson gave her slipped on her forefinger. Ashes always fell into the mix she made up of her paints, she'd eye whomever she had lying on this lounge and squeeze out fresh dabs of fresh oil paint round in a rainbow on the marble slab table.

She'd underpaint in sienna and cadmium white. Then do a sketch in terra cotta, to place the image, then she'd pour another bit of sherry in her glass. "If you only drink Bristol Cream," my mother said, "you're not an alcoholic." And, using burnt umber for the darker outlines, building the underpainting, with shadings of viridian green, molding and warming with venetian red, she'd comfort her models. When I said our governess was crying a lot, my mother painted her picture.

It's late-night lighting: I push the model's chair off stage right.

My mother wheels the easel off stage left.

She returns in a long satin negligee, pushing a door on a frame. She stands inside the frame, holding the door. I am sitting in a pink chair.

My mother explains Milly is "frustrated, so she's seeing Dr. Greenson for a while." I ask what "frustrated" means. My mother bites her forefinger. I pick at the silver roses painted on the pink brocade chair.

My mother is lying back and looking up at the face she hates splintered into rainbow fragments in the crystal chandelier above her bed.

Her bedroom defined *boudoir*. Like their generation of Hollywood couples, my parents had separate rooms. But they didn't do it, they said, for independence, but to be like the East Coast American aristocracy, the presidential families, who did it to be like the European aristocracy.

Frustration was the major issue of the fifties. Unless you'd read *Lady Chatterley's Lover* or Havelock Ellis, my mother said, you most likely had no idea about sex. It's a sad, new kind of frustration to read about women suing men for coming on to them. How captivated we were when Helen Gurley Brown first wrote about our sexual independence. We just couldn't wait to exercise our new freedom to relieve our frustration.

"Frustration," my mother explains, "means she isn't having sexual intercourse." That makes it sound so formal. You'd do whatever it was with your robe on over your stockings. Stockings were important. Men were crazy about stockings, but you had to keep the seams straight or they'd leave you. Chipped nails, messy hair, broken-out skin, all, absolutely all smells, especially there, would scare men off like skittish horses when they see snakes.

But my mother did say that sex was beautiful. I imagined it would be as fragrant and silky as her Juel Park chiffon nightgowns, and might carry with it the thrill I'd get when I'd hear her play thundering moves from some of the great Russian piano pieces.

I remember someone telling me my mother knew nothing about mothering. "She wasn't designed as a mother," I said, "she was designed as an artist."

On those long nights, she'd lean against her pillows, sipping her sherry. She'd drink a few glasses every afternoon at tea time when she'd come in from painting to go over the day's messages with Betty, her secretary. They'd both have Harvey's Bristol Cream with ice cubes. Later, my mother would switch decanters, kept filled all around the rooms she might be in. She'd tell me how great my wedding night would be. I'd get that shiver I got when I'd look at some men's loins during chariot movies.

Grooming was just the beginning, along with thank you notes. You couldn't hold on to anything without writing perfect and fast thank you notes. My father corrected mine. Even the good notes could be better if I said it like this. He could write, even with the gloves he wore because of this terrible rash on his hands. He'd slather them with mason cream every morning, then wrap them awkwardly in gauze, and pull on the large white cotton gloves. I got the rash, too, and wore gloves. It was after that my brother and I had to stop picking our nails. Joy didn't pick her nails. Jeb and I were told we'd get fungus. The Marines coming home from the tropics had fungus. We'd hear the big-bellied planes roaring in late at night. And then the ambulances in the distance, bringing the soldiers to the veterans' hospital off Sawtelle. Every month there were more white-crossed graves.

I'm in London. The writers have left, we're cleaning up. "You did that very well," Stuart says. "No one would have dreamed you had, once more, no idea who the hell they were when they arrived. But I could tell you had another seizure when I was there having dessert with you."

"Don't be gentle just because you hate having them here," I snap. "And these are not—those aren't . . ." My words stumble. It's late. I'm tired, the words are like marbles rolling around. I'm trying to catch them with thongs—no, that's not it—prawns, but shrimp don't fish. Images are off. Tongs. Maybe.

"Don't call them seizures," I'm angry, "they're petits mals." Like petits fours. Charming little things you have for tea. "I don't know why they happen—there's no pattern—what really happened was that I really can't deal with reality. You always say, 'What's the point?' This is the point, to keep adding to the group so I won't wind up reading to myself."

"No," he says, "so you won't wind up in a room by yourself where you'd have to write—not just talk about how you can't."

Stuart goes into his room with the charcoal walls and puts on a new CD Gerry Mulligan just sent him.

When there's trouble with us, it's safe to say I am probably reacting to my parents. That first act of my life is the hard disk. Whatever my memory has dropped, the walls of photographs, shelves of books and portfolios of letters at my fingertips bring it right back.

This may be the next Sunday. Debbie arrives with a bunch of pink roses. Debbie is small. Her skin is pale, her hair and eyes are dark, shadowed, damp, like a child's. She catches a look at the antique toys around the shelves and on the floor and she's okay.

I have reread the note her father sent me before she arrives. He wrote that it might have slipped my mind, but we met at the first gathering of the new Fulbright Commission. He told me his daughter, a writer, was going to study at Oxford the following year and, in the meantime, he knows she would love to talk to another writer about her novel.

"It's nice of you to see me," she says, braiding the fringe on her woolen muffler, "but I'm not sure I can do this."

"I'm not sure what this is," I say, "but you wouldn't be here if you didn't think so."

I remember when I gave my father the book I wrote about my second marriage and addiction. As a father he was horrified, but as a writer, he said, "Welcome."

I liked that. You have to shock your parents to make some kind of leap. But there's the other leap into becoming the parent, the audience—the teacher.

"I have my story here with me," Debbie says. She has this shy persistence I understand. "It's a difficult story," she adds, and looks away.

"About an older man?"

You can't write without sex coming into it (even by an obtrusive absence), and if you're young and female, it could well be an older man. How easier in our world to see it as his intrusion; how much more naturally that moves from the page to the reader than that forceful longing you might have had for a colorful older man in your life.

"An idle guess," I say, "so come in." I put my arm around her and explain that I do confuse things.

"My dad told me," she said, "and you told me yourself when I called you last week."

After everyone else leaves, Judith comes by with Noah for dinner.

"Did you talk to Jeremy today?" Judith looks me over.

"You know I didn't," I snap.

"So why don't you call?"

"Because I don't want to be like that, you know, where the phone rings and you know it's your mother."

"When was the last time I could have talked to our mother," I never called her mom, "do you know?"

And when, I could also say to my brother, Jeb, who has just answered his phone, was the last time I talked to you? But I'll leave that.

"You were there when Mom died," my brother says. He has an "I'm going to tell you a story" pace to his voice. He says hello, and you wait for it to begin.

Dad had that, but with him it was hello and the cut of the story was likely to be, "This is what you should be doing."

The hateful parental role.

"You were there for a week right up to the end with Mom in the hospital every day. Then you had to go back to London, two or three days before she died, which was two days before my birthday," my brother says, "and I was bugged with you, leaving. It was those years after Dad died that I got close to her. Do you remember the scene about his Oscar?"

It's like a set of one of my father's plays: the family scenes he wrote over and over, missing his own, catching the powerful mother, the renegade showy brother, the witty sister, and the earnest kid brother with the big dreams.

This was their apartment in New York on East 70th. Jeremy was organizing plans for the memorial service. My mother made it clear she wasn't speaking.

My father had not made a will. He knew he was dying. As he looked around their rooms in their new apartment, there was both too much to sort out and not enough. That was the hard part.

In the family tradition, he'd spread his money around, helping all of us when he was making a lot. Now, what was left were masses of books, even after the fire, china, crystal, and silver. Lots of bronze and silver engraved awards, prizes, and cups.

My brother tells me how it was.

Mother was sitting off to the side, her neck arched, jaw jutting up, her skin frosted crystal. She was aloof, almost contemptuous. My mother's emotions—her pain and her love—were unique.

My brother stood by the fireplace, watching as we cased the scene. The only thing we didn't want to deal with was Mother. Did my father ever think she'd outlast him? Did she think she would? Or, the question I ask, thinking of her last lonely years, hands crippled with arthritis, not painting, not near any of her children, did she outlast him? These years of my mother's life I don't want to remember. But I do now. They began that day, watching us eyeing the pieces of their lives as we talked about the memorial.

Joy was organizing speakers.

First I asked my mother, "When would you like to speak?"

"I don't want to," she said.

"But you have to."

"No," she said, "I'm not speaking."

Joy was tearful. I detest that Joy can cry. I envied her tears. I hated that I was in such control, sitting next to my mother, who shrugged me aside, keeping me from touching her.

Jeb said nothing at first, looking off out the window. After a long western pause, he said, "I'm not talking."

"But you're his son," Joy said. We knew how much that mattered. My mother used to say, after nearly dying giving birth to Jeb, "I'd rather have died than not give your father a son."

"But you have the name," I told my brother.

"I said I'm not. I don't want to parade around what I'm feeling."

"I felt," my brother tells me now, "I was out of line, but I said, 'it's not going to happen.'"

Mother looked at him. "I understand exactly."

"She saw I didn't want to share it," Jeb says. "'It's none of their business. You do what you want.' And suddenly," my brother says, "I thought, there's something here for me with this woman. And I'm going to give her my time. She didn't want to speak, didn't want to look at it. I should pay attention to this woman. There's more here than I know."

We talked over the Oscar. Joy thought she should have it for her son, Saul, who loved films. I said, "Jeremy should have it; he's in the business." And Jeb said, "The name Schary is on the Oscar. My name's Schary. It's my fucking Oscar."

"So," Stuart says when he reads this, "do you remember what you said at your father's memorial?"

"I didn't speak."

"You did. You talked about your mother. You stood at the mike and you talked right to her, about their attachment, about the support she gave your father, about how much you learned from her about love."

I cringe as he tells me. "I have a hard time remembering good things I've done."

But I see her sitting there, in black, her face like chalk, her expression stark. No matter what I said, the only thing that ever really mattered to my mother was gone.

"*I*f you made a list you wouldn't have to go back and get things," Judith points out. She has picked me up on Marylebone High Street after my second trip to the market this Sunday.

"But if I made a list," I explain, "I wouldn't be working my memory." This is one of my favorite strategies. It is not popular with people who have to get things done.

Dr. Barry Gordon has straightforward strategies, advice, and memory aids. He suggests notebooks. I tend to lose them or fill them with drawings and give them away.

I repeat, write down what I want to remember.

I put Post-it notes on things I have to remember so I'll notice them. Stuart's son Philip has told us a brilliant new thing is coming: a watch that has, he says, a global positioning system, so when I'm lost, a map comes up and tells me where I am. This will work if I've written down where I think I'm going.

Be consistent. I always do things in little routines so they become familiar. I brush my teeth with the red toothbrush in the morning and the lilac one at night. Hello. And good night.

I connect everyone to someone else. I try to remember people I really want to remember by setting up a ready-made memory theater: starring them in a favorite movie, putting them in a scene I'm wild about. But then I think of the part and not the person. Nothing works every time.

Barry also says, "The less you worry about memory lapses, the less likely you are to have them."

It's the connections of reflections, of ideas, of concerns that make the point. People I love who really remember me aren't thrown—it's one more quirk.

"The memory you really need can come naturally." Don't, as James Billington, the Librarian of Congress, says, confuse information with memory. The real memories are the emotional connections, like the wonderful song from *Gigi* that I can't remember when the two people once in love, long ago, talk of their times together. They each get the where and the when wrong, but they agree it was wonderful.

I tell stories, incidents and scenes I want to remember, over and over, fast. They will become part of my repertoire.

Learn in small bites.

Have a memory partner. I'm well covered. Susan Granger in Westport, Sandy's in Boston, Holly in New York, Joanne in L.A. Stuart is my memory partner in London.

Gordon also points out, most helpfully, that there may be things I don't have to remember.

I've been holding onto old plots, old addresses, and, really useless, old hostilities. Characters killed off in old novels. I write them all down. Tear them up and throw them away. Put them into scenes I cut. I think they'll disappear.

So I don't, before going on a trip, say, fill my head with packing. I draw all the things I need to pack, put the pictures on the changing room door, and when I think of packing, I can scratch them off. So I don't worry about leaving something behind, I put the list of pictures in the suitcase.

Paying attention: If I don't remember someone, it's often because I wasn't interested. The artist Andrea Tana introduced me to a lawyer years ago named Simonetta. She smokes a pipe. That made me pay attention.

Then I watched and listened to her. I have never forgotten her. (To catch attention, do something no one else does. They'll see you're memorable.)

Where that's all far too simple, consider what the poet Simonides accomplished 500 B.C.E. He'd gone to a party to do the toast celebrating an Olympic victory by a very popular wrestler in town. He left right after giving his poem, only moments before a landslide collapsed the hall. The dead guests were unrecognizable. Simonides had taught himself to fix faces in their setting, as parts of an entire image. So he remembered where everyone had been sitting and in that way identified the bodies. Cicero formalized this approach around the first century, and it was picked up again by the Russian Shereshevski.

For example, I am going shopping. Before I go out, I imagine the things I need really clearly in my mind placed on some particular spot in my dining room: fettuccini on the chandelier, Stilton cheese on the board on the table, foil on the silver tea caddy L. B. Mayer gave me for my first wedding—and so on.

This system works for me when I absolutely have to remember a few things, so much so that I do not throw in such distracting details as how I got the tea caddy and of how many ways Judith reminds me of L. B. Mayer and other moguls. Even on a simple day, like a Wednesday, I can only place and remember seven or eight things at a time.

Sunday is never a simple day for us.

And then, no day is simple since I have been trying to write again. The main thing that happens is angry nightmares about not writing that turn into petits mals. And the writing lies dead. I ask Stuart to listen as I read. My voice does not charm him into lying about the work.

"It's a nice idea," he says.

"This isn't where it will really go," I explain.

"I'm sure," he replies, "it's good—it's different." He smiles.

He tries. I miss my women friends, the network. I look over my friends' books from long ago. Josie had just published her book when she brought me up to her agent, Lynn Nesbit's, office. I think I was wearing the fifteen-dollar midi skirt I bought at Macy's. I take the books down from the workroom shelves, dusting them off, rereading. I remember sitting with Lois Gould in a restaurant, switching chapters we'd pull out of our bags, and I see a big room overlooking Central Park West. Was it Betty Friedan's or Alix Kate Shulman's? The entrance halls were full of roller skates.

We were with each other in the sixties. Or was it the seventies?

Life Signs—Josie had the perfect title. The world seemed ahead of us, we were stepping out into it, out of the isolated kitchens and away from the dreary corner desks into what felt like careers, jobs. Real writers.

Now writing feels like a conceit, I tell myself as I sit in front of a yellow legal pad in my London workroom, Victorian rooftops framing the gray sky. I don't have the smart pacing you get from having an editor down the line, like when I was writing for Helen Gurley Brown. She was a tough copy director, writing advertising copy for Foote, Cone and Belding in downtown L.A. Helen taught me how I could write ninety commercial radio shows for women stuck at home, thinking long and hard about laundry and what to feed the kids for supper; neither of which subjects came to our minds in a given day.

"Do you think I can pretend I am Helen teaching me?" I ask Stuart.

He is sitting at his gray birch tree desk, working. I have this knack for knowing when it is a particularly bad time to start talking about my writing. Always on a Sunday evening, after the writers are gone, when he's got a conference here on Monday morning and he's putting his notes together.

"You act as if you are Helen when you are working with someone else, sharp and organized." He wants me to be sharp and organized and let him get back to his own work.

I like making it easy for people who meet me by talking about the memory problem. That was what I loved about my mother; right away, if a friend of mine came over, she'd tell a fresh story of how she'd fallen or how it happened, an inspiration to me, particularly now. Not-remembering is like situps for the imagination.

This Sunday, Judith is here with Stuart, and Geraldine and Debbie are here with me.

Debbie always brings flowers. She has added this suspense to the group: can she keep her shy distance while turning out these dark, original pages? She writes I fear out of her own memory. She tells me today, when I am missing what I still call home, "Rilke suggested trouble may be guarding the gate to Paradise. Maybe what we miss is what we shouldn't have."

I pull the chicken, roasted golden with sweet paprika, out of the oven, and reach up and take down one of the colanders for draining the fettuccini. "If I learned cooking by watching my grandmother, why didn't I pick up piano by watching my other grandmother teach my cousin, Julius Katchen, to play the piano?" Different part of the memory.

I hand some red bandannas to Geraldine. I see that by piling them carefully according to pattern and shades, I can now remember how many I have, which restores my confidence on days when I forget what I did last night.

"Today I have remembered you don't eat mushrooms," I tell Debbie. I put her flowers in a pitcher and bring them into the dining room. But maybe it's Angie who doesn't.

I sit in the chair at the head of the table in our red dining room. Geraldine always sits on my left, facing the fireplace painted like a merry-go-round by my granddaughter Phoebe and me, looking out on stacks of books on the windowsills, hanging paper masks, mechanical toys, antique banks, carousels, and Shaker boxes. Angie usually sits on my right, opposite the picture of Justin in his Yankees cap and the pine buffet with its parade of tureens and the movie star plates Stuart found, balanced on both sides by Laurie's picture of us looking at each other and her portrait of me seated, writing, with a giant theater backdrop.

Mark sits next to Angie farther down the long oval table, which I've covered in large swatches of dark red Liberty prints to go along with the ones swagged up above the lace curtains on the front bay of windows overlooking the street. Mark is late. And one of the swags needs tacking up. Angie takes my hand. "I have met someone." She is very private, but she says "someone" as you say it when it's really someone.

Low early-winter light dances off the crystal chandeliers and filters in from the window overlooking the mews I see from my father's old Early American desk, where his brass student lamp with the emerald green shade lights up my family photographs.

Stuart's life-size plaster ram looks down from one of the big bookcases, all topped with deep thickets of dried leaves, branches of dark red, sienna, garnet, and plum; bunches of heather; and giant bright crepe paper poppies.

The knife, fork, and spoon from Schary Manor, my grandmother's catering place, are framed to my right. This week is more like my grandmother's Sundays, like my parents'—it's not just Geraldine, Mark, Debbie, and I passing around food and reading out what we've written this week, and then, all so carefully, telling each other what we think, the goal being more about staying with it than cutting it up, although I never exactly said it like that, but you look at any one of us and know that's the deal.

The writing group has lasted longer today because they all have something going in their lives besides writing. Mark's uneasy because Archer has flown off to Mexico City because there's supposed to be a great new treatment for him there.

"Archer's fine," he says. "Did you really like my pages or were they too long?"

"No, really, the pages are fine . . ."

"Archer is not fine," Geraldine says in a low, sympathetic voice.

"He's comfortable," Mark doesn't miss anything, "of course he's not fine. You're right." He turns back to me. "Are the pages only fine?"

"No, no, Mark, they're wonderful."

"But do you really remember them, or is that to make me feel better?"

"I take notes as I read, you know that—you can see."

Before Geraldine leaves, Stuart comes into the kitchen for coffee and Geraldine hugs him. "Sometimes you are both like parents to me."

But "like parents" is not parents. "Like children" is not children.

Have we made up surrogate families who see us as we wish to be— this ideal romantic couple? How do we look to the families we've left behind in America?

Johanna knows the best way to keep me in touch is not to remind me how much farther apart the distance puts us every year. We deal with it by our phone talks, creating a radio series relationship, all neat and chatty, with only a few teary scenes now and then.

Stuart reminds me that Jeremy sent me a ticket to visit him and to begin to know Phoebe. "He made it clear that means a lot to him—for you to be there." I was there, staying with Jeremy in L.A., but I don't remember that trip.

I have come into Stuart's room. It is the twin to the dining room, except it is dark gray, beige and neat. Judith wants to stay to talk to Stuart about her studio and the new ways in which she hates it this week. And Laurie has come upstairs to show him the contract her new gallery has sent over.

This is the kind of Sunday I love—and exactly the kind he hates.

Now the phone rings. It's Archer from Mexico City. "He wishes he could just be here," Mark says.

"Why not," Stuart says, "everyone else is here." Stuart takes the phone.

"He has this infuriating way of talking to people in front of you without letting you know what they're saying," Judith says.

"No wonder. He was a spy," I say

"When did you know that," Laurie says, "or did you just make it up?"

"He was," I say, "in Hungary."

"She doesn't forget what interests her," Judith says. "That panel over the fireplace is really terrible," she points out to me. "You'd have great stones under there if you took it off and put some tiles above . . ."

"Not this year." This year rent is the issue. He'll kill me if I told anyone, but by tomorrow he'll have me convinced we're fine.

"It's not stones," Laurie says, "I have the same fireplace. It's brick-work—hideous . . ."

"Oh, one of your valances in the dining room is falling down," Judith tells me.

"I know," I reply, "and my lace curtains need washing." Suddenly I remember myself telling my mother her silver wasn't being polished enough.

"I hate nets," Judith says. "That's what we call these lacy curtains over here, or did that slip your mind?"

"You're cranky today," I say, "or is this how you are every day and I've forgotten?"

"No," Judith says, "this is unique."

"Yeah."

"The nerve."

"The Cat's in the Cradle" thing is true. In this latter season of life, you do wind up paying for every bitchy, ratty, thoughtless moment. Tough if you can't remember them. The payback itself is usually enough to open the code. I snap my fingers and say, "Oh, this is for that."

Memory is a series of chips. You draw a card and you can jump a whole stack of chips, or throw the dice, never knowing where you'll wind up or in what piece of time. One great thing about being older than you thought you'd want to be is there's so much more to remember, to mix up and to put back together.

It does not surprise me to be in England. It's a cold pink-and-blue dawn. The rain hasn't come around yet.

When I was a child, one of Hollywood's main concerns as it became alert to the Nazi intent, before the rest of America, was to keep America interested in the war. The government and the military were aware of Hollywood's importance and our lives were filled with people from these worlds; the way the war caught our hearts was in the stories of England, the faces and voices of the English actors who came over as refugees. A lot of their children went to our school, and my favorite teachers were English. England was small, valiant, and had a courtly class system, just like home. London looks like several sets of itself and works like light comedy.

I'm sitting at my father's old desk, looking at pictures of him with Cary Grant. Here's one of them out driving. I call my brother in Texas. "Where were they driving together, Cary and Dad?"

"That was a scene from *The Bachelor and the Bobbysoxer*. Funny movie. Dad played Cary's friend in that scene."

I stay up all night, going over this last talk I had with Cary Grant about my dad, all wound around in notes, the way I do, half drawing, half writing.

I met with Cary Grant on a Bel Air patio, surrounded by camellia and hibiscus bushes.

I arrived five minutes early, so that I must wait—a procedural memory of protocol picked up watching costume movies about medieval courts that screenwriters copied from Hollywood's social scene. Cary, Spencer Tracy, Robert Ryan, Humphrey Bogart and Gene Kelly were probably the actors my father knew best and the ones who were closest to him, from his earliest days in Hollywood. Spencer Tracy starred in *Boys' Town*. My father won his Oscar for the screenplay.

My father went on to be a producer for L. B. Mayer's son-in-law, David Selznick, who was married to L. B.'s daughter, Irene. After my father left Selznick, he ran production at RKO so successfully L. B. asked him to come back to head up MGM's production. There, in a coup by Archer Schenck and the East Coast stockholders, my father was put in to replace L. B. Mayer, even though Selznick warned my father that running a studio which is not your own never makes anyone happy. The stockholders at the Board thought my father would be more manageable, and more economical.

Cary and I talk first about England, where Cary came from "with an acrobatic troupe in 1920. There was a modernity, an excitement to New York," he said. "I met your father first in New York at the St. Regis Hotel. He did a scene with me just before I left for the Coast. We met again in L.A. when Dore was writing for Metro. L. B. Mayer was a moralist and he hated writers, but I always insisted on writers being on the set. Directors never could get the timing right unless the writers were there. You could look great riding the boom with the pipe, but looking like you're directing and directing are two different things."

"Mayer pulled stunts," Cary remembers. "He'd get Dore on the intercom and give orders in a tone and a manner that indicated he had someone in his office, that Mayer wanted to show he was calling the shots. He'd never talk like that if he were alone. Your father told me, 'The son of a bitch has a roomful of people in there.'"

There is no bad light for Cary Grant. With the sun on his weathered tan skin, shadows shape even deeper charm into each expression, if such is possible.

"Dore was sexy," Cary says, "I should say, a sexual man. But, first of all, your father was crazy about your mother; she was the guardian demon. Miriam was the only person I ever had a relationship with exactly on her terms."

"I'm not sure I want to see my father with anything approaching what people call reality," I say. "I don't want to detach from him, to view him with dispassion, or to put him in perspective. I'm not interested in taking him apart. So I'm not current—I won't excavate my parents' lives to answer my own questions."

"I don't think you mean excavate," Cary says.

"No, but you know what I mean. Look, I'm sure there's no more grief in my father's story than in any story of any creative, disarming man's rise to that kind of unmanageable power, but people expect you to come up with dark stories."

"Exactly," he says. "A lot of people in this town resented his goodness. He intimidated some people. He had an almost religious respect for academics. He'd fill the house with academics to demonstrate he was of a studious turn of mind, but they'd be bored to find each other there—'Where was Lana Turner?'"

Or, which he doesn't say, "Where was Cary Grant?"

My father had the star's way of walking into the room and being the center of it. This makes a lot of people in L.A. hate you automatically, but they know they have to handle it, not fight it or ignore it, or they'll lose out.

It's a particularly difficult thing in a large family. Some other people aren't going to like you for this, but each gathering you do show up for lights up more for your being there. Our family had its share of "A list" types. Their most disastrous impact is on brothers-in-law, or actually any people who marry into the family without taking the status game seriously.

The outsiders, the philosophers, rabbis, and psychoanalysts who came for deli dinners, watched my father as closely as the movies he'd show, as curious about his idiosyncrasy in Hollywood as he was about them. He'd watch them talk, and I wonder if he was saying somewhere to his own father's critical spirit, "See where I am? See, they even listen to me."

"He wanted to be considered the innately educated man. His lack of education irritated him," Cary continued. "He enjoyed giving the academics he had over what they didn't have—and then their education might rub off.

"Dore never saw himself as an employee. But then, L. B. treated Dore like his idea of a son."

His idea, maybe—but then, I doubt if my father knew what a son was supposed to be. His own father left his mother and went somewhere when his business failed—I have the impression it was England. I stop to wonder if my kids see me like that.

My grandfather never sent anything back home except an occasional wild piece of Victorian china. I remember (and suddenly I really do) a giant punch bowl supported by six naked cherubs with garlands of flowers. Three feet across. I loved the punch bowl. The cherubs were naked and "uncircumcised," my mother said with a sexy look.

My father walked sharp through the MGM back lot, maintaining his gait the way he taught us to walk on our hikes along Tigertail. He surveyed his sets like L. B., with assistants keeping up around him and the studio photographer sprinting on ahead, backwards, snapping every exchange. This authority and fast delegation didn't come easily for my father. He liked the reflective pace and shifting concerns of a writer. This is why other writers didn't run studios, and why my father spent a lot of this time in bed on his back.

He had leadership he liked to use; ideal for becoming a producer, as Gene Kelly said, "He was a born executive, not a born writer. Eyes gravitated to him."

Cary says, "I think Dore preferred the idea of being a writer more than actually writing."

"But that," I say, "may be the first symptom of the writing problem." I don't remember how, but I do remember I don't like it and must do it.

"Do you think, as someone said, my father was the ultimate naif?" I ask Cary.

"No," Cary holds his hands up together, fingers touching the way my father used to do when he was thinking something over carefully, "I think Dore found himself in situations he didn't understand because he was either unequal to them, or incapable of impugning any wickedness to other people's motives. Your father was the audience, no less dazzled. He didn't do it for acknowledgment; he believed it. He wanted recognition—we all do. Perhaps, more than anything, Dore wanted congratulations."

We talked about the Sunday nights and the other nights when MGM's head editor, Margaret Booth, came over. Cary was one of the few actors who was close enough so he could just drop over to watch his own scenes.

"Your father started running films at home, and a darn good idea, too. Dore was modern, but he may have had too much imagination to be a mogul. Moguls deal in economics; not that writers or actors can't. Why divorce the actor from the businessman? If you enjoy acting, you'll do well with it. Dore had more of that same knowledge, that streetwise sense you get from actors.

"You know as an actor when someone's real with you. Of the best actresses I had the good fortune to ask for, the best was Grace Kelly. Hitchcock thought the same thing—there was a contact, Grace could listen. Most actresses aren't fully listening; they're wondering how they look. But Irene Dunne had great comic timing, and Betsy [Cary's third wife, Betsy Drake] wrote dialogue very well."

You become so much more conscious of World War II when you live in England, and you're more grateful and slightly protective about being Jewish. We go to Corinne Laurie's for a Sabbath dinner every Friday in London.

I met Corinne at Andrea Tana's, the American painter who has created a Mediterranean garden in Kensington. Andrea knows everyone interesting in London.

I was grabbed right away by Corinne's dry humor and flair. She whizzes around London and the whole world with such snappy style, you'd never guess the grave losses she's had.

Stuart says we have been picking up Corinne's father, Max, for a couple years to go to Sabbath dinner. People our age may be considerate to people who are really old because we're just beginning to see we won't be able to escape it. What I really feel in these months of Fridays is that I'm an expatriate—like Max. Except my family is not dead. I sometimes say being here is the best thing I can do for them. Do they want to deal with my anxieties, fears, and demands, my needy ways and yawning frustrations, which I cover by cooking, drawing, and changing clothes? If I feel this distance from my land, what must Max feel?

When I was young, I never wanted to be near old people. I might pick up mannerisms, fading attention, less snazzy language. Words like "snazzy."

We were stuck in traffic this one Friday and it occurred to me to ask Max if he left Germany before the war. "No," he tells me, just like this, "I was in Germany throughout the war. It was a terrible time, and to be part of a nation that is doing something you hate. My family went to the camps. They died in Buchenwald." I am wrestling to hold onto moments so I'll have a past—while his whole life is really gone. His gears seem stuck at times. He repeats and repeats a sentence. I stop. Do I do that? Max is living in a recurring cyclorama, old films going around in the head day and night, the suction getting stronger. He shakes his head sometimes, over and over, as if he's stuck in a thick, terrible place. The wheels spin, and I hate myself for getting impatient.

It's too close for comfort. Closer for comfort is when I'm watching Corinne with her grandchildren. As they bring out the family albums, talking together, passing 'round the newest baby, laughing over the pictures, I catch what I've ripped from my children. I am their memory, and as my own memory is gone, so I've taken theirs.

My father's own father told him he'd ruined his business after some of my dad's friends showed up and sabotaged the first catering event my grandfather ever let him host for him. In reading and talking with my cousins, I realize the catering place fell apart in the Depression.

How much of what we accomplish stirs our parents' ghosts? Maybe their disappointment in us isn't the real issue, and what we're remembering are their own big dreams that blew up in their faces.

I spent my school years studying my parents and their world more than anything else. But now, when I consider my father's fastidiousness, which I used to lay on his being a Virgo, I'm sure it had far more to do with the influenza epidemic around the time of World War I. The way he described how it was, in just his house alone, makes me snappish when people call me now and say they've got the flu, when all they've got is a cold.

People were sick on all five floors, my father told us. His mother, weak and drenched with fever, was clutching her new baby, running down the stairs from her room. My father was twelve; he knew that the baby was already dead. Plumbing had frozen. He was surrounded by the stinks and sights and sounds of sickness.

He told us the story when we were kids. And I colored it up for story time at school. I said that all of my father's brothers and sisters died during the flu epidemic and he became a writer after he wrote about them for his mother so she wouldn't forget. Perhaps, I thought, reflecting on my father as he was during the years when Jeb was always desperately ill with asthma, my father wanted to be sure that in her grief, his mother didn't forget she still had him.

The principal of our school, Mrs. Dye, who had lost her own son during the war, called my father to tell him how well I'd done and to talk about what a shattering thing that must have been for his mother to lose so many children at once. My father gently explained to me how important it is to separate what really happens from the story you write and to be careful with other people's real lives; to keep memory and storytelling on separate shelves in your mind, like reference books and novels.

I remember, as I read this, asking Cary if it was true what Ralph Greenson said about my father, that "he was defeated by his own goodness."

"There wasn't anything devious about him. People were accustomed to evasion—it almost embarrassed them. Dore recognized integrity, but he often failed to recognize its absence. That," Cary said carefully, "is how I would say it."

16

"Do we fight every Sunday?"

"Just about," he says. "I like to be alone and quiet. You like to have as many kids over as possible."

Last Sunday night I had Tim and Lucy Mellors and their three children for dinner. The kids painted on the walls. I made my grandmother's pot roast for supper, mashed juniper berries for the jelly, latkes, but I did not avoid thinking of Jeremy and Johanna—not for a second.

They liked the dessert of crème fraîche mixed with strawberries, chopped almonds fried up into a kind of toffee with butter and brown sugar, and croutons of pound cake. This used to be severely good—but now it's kind of like eating Sara Lee shoulder pads. Did I cook all of this for the Mellors or for Joanne and Gil Segel? In Santa Monica Canyon? It's best not to talk about it. He knows I invite kids because I miss mine.

He puts down *The Financial Times*. "This isn't a fight. This is how we are. We knew right away there would be times when we'd want to be on our own. I'd want to go hear jazz. You'd want to spend evenings with your kids."

"Did I want to spend evenings in the city and you want to be with the kids? Your kids?"

When Stuart's daughter first came to visit after we told her we were getting married, Susan was as anxious and uneasy as Johanna, but she hadn't the sharp cover Johanna had invented. One day we went to the City together. When families join, it's a kind of ballet; everyone has to do a *pas de deux* with everyone else before being able to join up for the *corps de ballet* finale. Susan was a kid I could never have raised. She has the gentle assurance of a child raised with order and discipline. She follows me around Bloomingdale's peering at me, not the merchandise.

"So why did you decide to be a writer?" Susan asks me. We are having coffee in a cafe on Lexington. The shopping didn't interest her much after all. This made me regard her with a new and intense curiosity. It could be because she was born in Ohio—some women born in the Midwest become serious and live meaningful lives because they do not actually like to shop. A friend of mine said that staying out of stores was a shortcut to excellent sobriety, one she did not expect me to master.

Has Stuart spent time with my kids?

Don't get into this. It is prime fight software. I can feel myself storing it away, saving it for another day.

This Sunday we will go to the movies.

We have been wanting to see a movie called *The Ice Storm* by Ang Lee. We say we want to see it because it's set in Connecticut and because Ang Lee is a great director. As much as I say I want to see this picture, I also avoid it.

So, it's Sunday afternoon. "We can catch the six o'clock in Leicester Square," he says. He looks at his watch. "We'll have to leave by five-thirty."

Our other fight is the being on time fight, a variation that assumes real significance when it involves catching a plane. I've explained this to him. And, as a compromise he doesn't really appreciate, I always let him have an easy win on the plane timing. I am ready at the time he tells me we need to leave, "and this is cutting it close," he'll add. This means we have two hours to kill at the airport, during which we do not speak to each other because the English don't fight in public.

I try, however, to win such other leaving-the-house games, like going to movies in time to see the commercials. They're a fast way to catch up on the best and worst new ideas. Leave too early and there's no suspense about getting there.

I like to go out to see movies, to catch the crowd reaction; as if it's my responsibility to pick up where they get restless, whether the laugh scenes go over. Some of the times, when I was a kid, we had to stay home and watch movies in each other's projection rooms because of kidnap threats from whatever radical political group our fathers' pictures had most recently offended. (This is why they made a lot of musicals.) Some parents would not have noticed if the kids were gone. The governesses and chauffeurs also liked us to gather in projection rooms so they didn't have to hang around with the actual public. I loved to feel the public's reaction to a movie. It was completely different from a projection room or preview reaction, and you could tell if your work really came off. When I was around eight and writing my first short stories, I'd go downstairs to Dottie's room where Ethel and Mabel used to hang out at night and read aloud. They'd tell me what they really thought. Sometimes they'd sit there. They'd rather be listening to the radio and the characters they knew better than the families they'd left behind in the East or the South.

"I hate seeing commercials. They shouldn't have them in movies. Do I need a raincoat?" Of course I need a raincoat.

The phone rings.

"You haven't got time for that." He looks at his watch.

"I'll just answer. Not talk."

He sits down and crosses his ankles.

"Hi, it's Steve. Steve Lewin. We said we'd call when we got here." He has a New York accent. "So we're calling." West Side. Steve Lewin. This could be the name of Erica Jong's new husband. Erica and Steve. A possible combination. "You told us about that little French restaurant you love—we'd like to take you for lunch."

"Sure." I've been in London a lot of years and I love seeing anyone from home. L.A. is really home, but New York's a close second. I lived in New York for a long time before I came to London. It's been three acts: L.A., New York, and London. Actually four. There was also Connecticut. The only difference between really living in New York and Connecticut is when you live in Connecticut, you still say you live in New York—you just have this house in Connecticut.

This man who I've decided is Erica's husband and I agree to meet at Villandry at 1:00 on Thursday. I hang up the phone. Put on the answer thing.

"I'll just call Danielle and make a reservation."

"Do you want to go?" Stuart asks, patience thin. "You don't have to worry about the commercials now."

I was working at Saks when I met Erica. She was small, round and uneasy, with long blonde curls like a girl in a Kate Greenaway illustration. "I'm doing a poetry reading at the Y," she said, "and I need a dress."

"This is not where you need to be." I pulled her aside and sent her over to Norma Kamali, a new young designer I was crazy about.

Erica invited me to her reading and I said I'd written a couple of books. She sent me an application for a grant from the National Endowment for the Arts, which made it possible for me to leave my job and begin writing again.

I'll read some of Erica's last book again before we have lunch—where did I put it? Remembering a line brings the writer, the friend, to life again—for a moment. Erica once told me, "You do not die from love, you only wish you did." She wrote a poem from that remark. Most writers never exactly talk to you. We test lines. You are material, you become a character. Your stories are our rough drafts. I loved living around writers in New York. I worked to keep my dialogue sharp. There was a better chance I could be someone else's character and not have to worry about getting hold of a story or an image long enough to make it live on my own pages.

"Maybe," I'm saying in the taxi to Leicester Square, "I am putting it off because it will remind me of Connecticut." I'm looking out the window of the taxi. "This is a great section of town."

"This is Soho. You were here last week visiting Judith at her office. Maybe," he adds, "you don't want to see *Ice Storm* because you're afraid it will remind you of nothing."

This is true. I don't have to say it. I grip his hand. So much of that last time living with my children has great silent, empty gaps. For my kids, the gaps are filled with people barely out of adolescence, left in charge while I was writing, taking better care of the books than I was of them. If you were old enough to drive, you were old enough to take care of yourself, was my idea—kind of a medieval concept of minimalist child care.

As I watched shots of the dry winter forest of birches, as fractured as the images glittering later through the ice-crusted branches, I saw pieces of those years, images of my children's faces, sunny, laughing, in their cars, at their parties, with their friends, reading, sleeping, angry. How many of those images are from snapshots? How much is gone because of my memory problem, and how much because I can't bear to see how it really was? I catch their anger and disappointment, but in my own voice. I can't always catch their voices.

In *The Ice Storm* there was a seventies key party (which I never heard about in Connecticut), where the women would throw their car keys onto the coffee table and take home whichever man picked them up. This was not a practical idea in Connecticut, because most of the cars would have to be jump-started from the cold. What you wanted was an auto mechanic.

I thought Connecticut marriages would be difficult and resigned, like John Cheever stories. "We don't play Westport," Susan Lardner said, "we're too depressed." Susan and Fred Hellerman, like most of the couples we knew, exchanged dialogue like they were practicing for *New Yorker* cartoons. These were the conversations to master; like duels, you slashed each other with lines about other people, movies, manners, and, above all, how they'd dress, move, and speak. Style was the thing to master.

But in how you decorated your house, not in your clothes. "What are you wearing?" was not a question. "Have you heard from the kids?" was interesting, especially when the kids were six and seven. And always, "What are you making for dinner?" "Can you use an orange quilt as a tablecloth with hydrangeas?" These were the questions we were asking Martha, Andy Stewart's wife, who was not having an easy time, even though she was working and running her house perfectly. You knew she'd rather just stay home and make it wonderful. Martha lived in Greens Farms, open, easy to find, with lots of land.

Erica's house was not so easy to find, Stuart pointed out the first time we went over there together. You're never fond of each other's friends' houses at first. How, you wonder, do I fit into this picture? Not easily.

Erica and her then-husband, Jonathan Fast, were living in Weston then in Erica's big, modern, perfect Connecticut house set on a hillside in a net of trees.

I try to get pointers on confidence by listening to Erica. When she asked Stuart what he was doing, she was the novelist, observing her dialogue and moves. "Now I am leaning over, asking him what he thinks. I toss back my thick blonde hair and laugh. I have a deep-throated laugh and he's enticed."

Sometimes Erica talks like her own biographer—some sentences have footnotes. "I'm doing a speech in the city next week; very much like the one I did in '79 at the New School. It was reprinted in the author's bulletin as well as my graduate newsletter. They published it with a great photograph."

Jonathan's father, Howard Fast, was there. (Howard was blacklisted in the fifties, which makes him something of a hero.) Stuart, interested, asked Howard if he felt frustrated as a former Communist watching America become increasingly polarized.

"I don't think that's true," Howard said. He picked at his food. "The working man has more advantages than he's ever had. I was over at Fleetwood Cadillac last week—any man can go in there and pick up a Cadillac at a very reasonable price. I haven't noticed it's gone up much beyond what it was five years ago."

"Jonathan's very cold and remote," Stuart said, "and Erica can't cook. But she's warm, I like her. Perhaps she can't cook because she really likes to write and she actually does it all the time."

This is something I can't understand. Erica lies awake at night and thinks about her writing—making it work for her. I lie awake at night and think about whether you could make a good jam out of juniper berries. I know I can make a jam, even an interesting jam.

Do most people think like this? Is having an actual memory like being saddled and bridled, or running on a track?

When I drove (am I more upset about not being able to drive than not being able to write?), I liked to drive on Mulholland or way up in Topanga finding dirt road connectors to Mandeville. I used to do rides like that with the kids in Connecticut. We'd just take off, winding way up in New York State, and come back down 59 over near Weston, near Erica's.

Her first husband, Alan, seemed cold and remote to me. Because he was Chinese? Culture is always a good excuse for difficult distinctions. Stuart is more private, more formal; "I don't believe in telling everyone everything." That's English. The land has a lot to do with it. No wonder Americans are open—we've got all this land, and some of it's warm and easy to live in.

I don't suppose, if you asked him, Alan would say that he was just being a stereotype. It wasn't the same with Jonathan. You looked at him and expected Woody Allen.

"He needs to be Woody Allen," Stuart said.

I'm in my changing room, dressing for lunch with the Lewins. I've changed clothes three times. I like American friends to think I am different now. So I won't wear the American frontier-look suede cowboy jacket everyone around here is used to. I put on my Sonia Rykiel pants and sweater.

I'm walking along Weymouth now, which still has the look of Regency England. I'm trying to remember which of Erica's husbands this is. There was Chip Wheat (a fine name for a hero of a story about a cracker). Chip was sweet, but rather in over his head.

Steve Lewin. He could be a doctor, or a dentist, like Fickling.

It is pouring by the time I cross Wimpole Street. It will stop when I arrive at the restaurant. England is not an island, but a small planet around which clouds circle like a flock of slow planes, dropping rain wherever I happen to be walking.

As a Californian, it never occurred to me to walk anywhere until I moved to New York. Then I recognized you could avoid the walking problem by moving to Connecticut. The English do walk everywhere, but London suits me; I can sit on park benches or in coffee places, framing people into incidents. Quick studies with a beginning, middle, and end. Nothing to connect. No continuity, no yesterday, and no tomorrow. There's only today. And most days there's only right now.

My house has a Wimpole Street address, but it sits on Weymouth Street, which goes into Marylebone High Street just below Villandry, which opened a couple of years ago. A tiny food shop/restaurant, crammed with nests of herbs, of mache lettuce, fresh breads in warm linen towels, challah, shelves of jams, cheeses, and sausages packed as tight as the people waiting for tables. It's a triumph. Now the owners, Jean

Charles and his wife, Roz, want to expand. They're bidding for the huge abandoned stable across the road where the Queen's guard used to keep her spare horses.

A small, dark couple out of a Truffaut movie, with wide nervous smiles, Jean Charles is from Paris, Roz is English. They were designers and decided it would be more secure to have a restaurant. We share the same sense of logic. I decided I'd rather be a writer than an artist because I didn't want to be working alone all day. My mother says you never call yourself an artist. Someone else will decide if it's art. My mother doesn't think much of it was. I think I've told her I see Clement Freud at Sagne's every morning where I have coffee. My mother likes his work. I must tell her he looks very much like the cream puff he has.

I had lunch at Villandry a week or so ago with Laurie just after she came up to say her mother died. She was sobbing and I held her in my arms. "It will be okay," I said, "and you forget, you didn't get along."

"It's you who forgets," she reminds me. "You didn't get along with your mother. I did."

The night before Laurie left for her mother's funeral, I was trying to remember if my own mother was dead and if I made up with her before she died.

I am thinking about this as I stand at the stop light.

Erica's mother is a painter, too, and probably alive. Mothers who thrive on conflict live longer. The ones who can say right out "You'll be sorry when I'm dead" don't need to die. My way to deal with this: don't write about it, just write Johanna a letter. Can I learn to be with her on terms not entirely my own? It might help if I was there.

Even though I live in London, I'm probably closer now to my mother in some ways than I've ever been. If I were going to be able to support my kids, I'd need power and a kind of shrewd, speedy toughness I don't see in my mother. My father had it. I tried to stay close to him so I could learn to be like him, not disturbed and fragile, the way I see my mother. But then I was too immature, too crazy, to realize that she had the guts to let me know her as an artist, real, varied, and conflicted, in a way parents don't usually show themselves. I hated her for doing that. I wanted her to just be my mom—consistent and not too close.

My mother is in me in all her ages and images. I miss her terribly. As I wander down the street and pass the windows of estate agents, stationers, and beauty parlors, I see my mother's paintings in an invisible gallery: portraits of the trees in our Brentwood garden, of movie stars in their costumes, of screenwriters, and of the Cape Cod seashore during the last happy summers when my father was writing his plays and they were alone together.

I couldn't have remembered those summers for my mother in exactly that way until now. At first I cast those summers in the fury I felt that they'd sold the house I grew up in. Now I see they had come back to the territory of coaching in summer camps, of theater—the territory of their new romance. The old memories are fixed from the stories they'd tell at dinner during visits from Moss and Kitty Hart.

I arrive at the restaurant and it's not Villandry at all. It's dark olive green enamel; and, like a nightmare room, it's almost empty.

I've done this before, I think. I walk up the street slowly to Daunt Books, a store of travel books built like a giant galleon, each deck devoted to books about a different continent. London is a perfect place to be if you don't remember people. I say, in my quiet London voice, "I seem rather confused." This doesn't seem to surprise them. "Could you tell me if Villandry has closed?"

"Oh, it moved—a few months ago now—over on Great Portland." He checks with someone else in the shop, perhaps to mellow any embarrassment I might feel. "You just walk along Weymouth and you'll run into it. Shall we ring Villandry and say you'll be along shortly?"

"Lovely," I say.

I go back to Weymouth Street, then across Portland Place, which comes down from the Regent's Park Gate, down past the Architectural College, which has its own Pâtisserie Valerie, which is where I will take my mother for lunch if she comes to London. I am planning all of this to make a deal with God: if I'm terrific with my mother, Jeremy will be great with me when I come to L.A.

As I cross the broad road into the new Villandry, I pick up my pace so it appears I'm pressed for time—as if I've just dashed over. I walk into the new restaurant through an enclosed forecourt like a French street market set in a huge art gallery. Baskets of miniature potatoes, morels, pink eggplant, baker's racks of warm olive and European sourdoughs, of challahs crusty with sesame seeds and marble counters spread out with platters of food to take home. You'd come here, I think, mainly to walk through this market. Then I do sort of a Kramer screech-to-a-halt entrance and scan the scene. But Erica is not here. I am late, of course, because I got lost, but they would have waited. And Daunt's did ring—but it only just occurred to me that they didn't have to ask me for my name.

There's no one here I remember I know. Then I spot a couple looking up and waving. Americans who have been here a while like to think we can pass as English, but we can't. We talk bigger, use larger gestures, are more animated. The English do not look up when you come in. The smallish man is standing. He is a potential Steve Lewin. Erica's husbands have been slight. The woman here, although possibly an Erica, is not that Erica Jong. But the tweeds guarantee they are Americans, dressing the way we do when we visit England. Unless we're of the later West Coast school, which wears fitness black everywhere.

"Hi," the husband says, "I'm so glad you could meet us. Great place." He looks around the restaurant.

"They only put the lamps up last week," I say. Since this couple knows me, I can pretend that I know who they are until we hit an impasse and it becomes clear to them I've lost my mind. "The architect is married to a friend of ours," I say.

"You told us last summer," Steve says. "His wife was just moving back to Yale to complete a paper."

"Oh, really?" I say. Would they go on, would they fill me in?

"Were you able to keep your house?" he asks.

"We're wrestling with that right now," I say. When you tell everyone everything, it helps you stay in touch with yourself at all times.

"And how's Laurie?" they ask. So they know Laurie. They must have bought one of her pictures. Yes. That's certainly it.

"Did you call Laurie, too?" I ask the Lewins.

"Yes, we thought we'd look at her new pictures, but she wasn't in. We left a message."

Jean Charles comes by the table. I still don't know Mrs. Lewin's name. I'll introduce them in fast French. Americans assume if you live in London you pick up French just like that. They don't know that the French believe the English have no grasp of any spoken language. The French use sign language in London, as the English do in L.A. Jean Charles tells us three of the lunch dishes aren't available today; he shrugs and smiles. Chaos is one of Villandry's distinctions. "We have ninety-three reservations and, as you see, we serve only fifty-two, so," he shrugs, "you can see how this is for the chef just now. Roz walked out," he shrugs, flings out his hands and laughs. Roz, the chef, is his wife.

"Should I come to the kitchen and help?" I laugh. This is a perfect way out. "I am so sorry," I will say. "I must cook." That way, the Lewins will think I am a little crazy, but have a talent for cooking known all over London. Not simply that I have no memory.

"No, no," Jean Charles puts his hand on my shoulder, "it is not necessary."

"Why would you take ninety reservations?" Lewin's wife asks, putting her glasses on to consider the menu.

"These things, they happen," Jean Charles explains.

This is excellent. We have a subject to discuss for the five minutes after he has disappeared into the kitchen—a subject that doesn't need any precedent. "It's difficult to figure how many cancellations you'll have, and at lunchtime people usually eat faster. But, here, they stay longer. There's more time spent on puddings."

"Puddings?"

"That's what we call all desserts here." I love saying "we" to Easterners. I'd never say it to a Londoner, and probably not to a Californian.

By the time we have finished the little onion tart appetizers, I look at them. "Listen, this is crazy," I tell them, "I don't know who you are."

"Jill," she laughs, "We met last summer in the Hamptons," They will dine out on this lunch.

But then, so will I. Except I may not remember it by the time I get home. "You see, I have a memory problem."

"Oh, we all have trouble with that," he says, maybe worse than forgetting is fearing you are not memorable.

"I feel really awkward here," I say as I look from one to the other. I case her crisp red hair, freckles, rosy cheeks. Up front authority in the fast smile; the eyes work the room with the wily routing of a London Courier bike. "But I don't know who you are—and," I add quickly, "that's not your fault. You'll have to tell me how we know each other." I look from her to him, quickly dipping ciabata into olive oil. "You see, this thing happened—when was it—a few years ago perhaps—and I really have lost my memory."

"So, how's Erica?" he asks when I come home. He's got a meeting on, so I simply say, "Fine. She sends love."

"Sometimes I'm tired of making it funny," I tell him later. "I remembered our first bad fight in Connecticut."

"Really?"

"It was about Sunday. I wanted to have everyone over. You wanted quiet."

"Yes, I did."

And now, too.

In Connecticut, I stormed out of the main little ranch house where he was working, out into the snow, across the yard to the studio house with the bed where we slept. I stamped up the steps to my desk in the loft.

He had called his friend Townie, who listened for a minute, then just said, "You want to stay married?"

"Sure."

"Then do the opposite of what you feel like."

So he took off all his clothes, crunched through the snow in his boots, pulled open my door, and stood there and shouted, "Let's make love!"

And I shouted back, "Fine!"

"Good idea," he says now.

C·H·A·P·T·E·R

17

This is now. We are watching a television show called "Why Men Don't Iron," which explains the radical difference between the way men and women think. Women think verbally. Men learn in action in images, which is why, for example, men are generally better at physics. Women tell you what they think. Men don't. This explains to me, once again and rather better than I've seen it done before, why it is that Stuart doesn't talk about his family and rarely sees them, but really cares about them and thinks about them as much as he says he does. Stuart rarely misses a day ringing his mom, even if they are on the phone no more than a minute—Stuart listens while Hylda tries to find a ramble intriguing enough to keep him on the line, to keep the sound of his breathing there a bit longer. Mothers of sons will get what I am saying here.

Yorkshire life is about work. This is why I understand Yorkshire, where Stuart grew up, better than London.

They once made licorice candy in Pontefract, his hometown, but it's mostly about coal mining. It's much easier to be in a one-craft, one-industry town. Even if you meet someone new, you have a hook for conversation. You fairly well know they'll talk about what happened at work today, and there's always going to be something disastrous in Pontefract.

The community is about work, like Hollywood and New York. In London, society seems to operate outside work. We don't go out with people we work with, and you don't ask people you socialize with what they do until you've known each other quite a while. Work and "society" are separate, exclusive.

Stuart is his mother's star—he's everything she waits for, the point of living this month, this year. His voice on the phone is the light of her day. He calls on Thursday to tell her he'll be there, Sunday, for lunch.

Sunday's her birthday. Mine is the day before. I read in his journals that we were in Pontefract for her birthday last year, too. Last year, he wrote, he gave me the Zodiac scarf. We gave his mother Elizabeth Taylor perfume in a purple satin bag, and a rhinestone heart locket with a matching pin. Stuart had written, "She's eighty-four years old and still has all her silver hair, so I'll have mine, too!" He wrote about talking with his brother and sisters about what to do when their mother got older. "I won't have her sitting in a home, surrounded by people who have forgotten their lives, staring at each other, not having a clue who they are or who they might have been." He couldn't imagine anything worse.

Hylda, his mother, calls on Friday evening to say we never came. So he reminds her we won't be there until Sunday. And this is only Friday.

Then Hylda calls on Saturday morning to see when we are leaving, and doesn't exactly remember who I am.

"I'm not certain myself," I say.

"Can I speak with my Stu's wife then?"

I know she means Margaret. The mother of the children remains the real wife.

Stuart's driving the Jaguar himself to Yorkshire. Mark brought it around from the garage up the street and washed it for Stuart before he took off to the hospital to visit his partner, Archer. Lilia has made a plate of brownies for Archer and a ginger cake for Stuart's mother before she left for her French lesson last night. She also gives me a birthday present, "not to open until the day," it says on the card where she's drawn her own coyotes. I ask Lilia when was the last time she went home to see her mother. "A year and a half. Maybe two years. My sister is the one who wants to be there. So she stays there. I want to be independent." She stops for a moment, then says, "You will be back on Tuesday and we will fix the plants. This is for real?"

"Absolutely," I say, "don't worry."

She stands, watching, waving as we leave.

There is no speed limit on English freeways. Stuart starts at eighty and moves up from there.

Last night I read his journals again to get his family's names right. He writes that he has secretly bought my birthday presents. By the time we leave, I have forgotten what they are.

Hylda will be calling again to see why he isn't there yet. So he asks me to call from the car to say we're on the way.

"On the way where?" she asks.

"To see you," I answer, "I'm calling for your Stu"—get the jargon right so I don't sound so foreign to her, although they don't mind the western American accent in Yorkshire so much. They remember it from cowboy movies. So do I. And I use it.

"Is Stuart coming today?" she asks. She'd have been up at dawn this morning, worried sick he might not be coming, wondering to Marlene, Stuart's sister, how he'd be out so late and gone so early.

He reminds me we always drive up to Scarborough on the coast the day before we visit his family inland, across the moors. Scarborough's like old Santa Monica. We walk along the boardwalk in the moonlight on Saturday evening, eating fish and chips. You need to eat them wrapped in a newspaper to catch the point.

So I understand Yorkshire. But I didn't understand Stuart and his relationship to his family.

Stuart's mother lives like a salty little queen in her cottage on a hill-side nest of a village outside Pontefract, whipped by the mixed air of heather, coal, and coming rain. Perhaps when you're old, memory's loss is a blessing. I'm not certain. Hylda's working at being cheerful, but as Sunday wears on she's increasingly lost and sits fingering her handker-chief, watching her family seated around her. The family doesn't visit a lot and they finish what they have to say in a minute or two.

I always start talking and keep at it until I find what I mean to say. But a Yorkshire person decides the one thing he needs to say before he sees you, saves it for the best moment, and says it as fast as possible. Once when we stopped to buy his mother Pomfret cakes and Allsorts on a visit, the sign on the shop's door just said SHUT. When Holbeck Hall, the hotel we loved to stay in fell into the sea, a notice on what remained of the cliff said "GONE."

On these holiday visits, some of us stare at the heater as if it's the fire-place that used to be there, while the men stare at the TV sports until after lunch, when his brother, Philip, comes up from the mines to go with his brothers-in-law and nephews to watch the TV sport at the pub. It has taken a while for me to catch on that Philip the brother is not Philip, Stuart's son, who is blond, lives in Portland, Oregon, and has a baby, Tucker, who lives somewhere else.

Stuart's sister, Lynn, pitched a bucket of water on Philip once and Hylda winked at Stuart and threw him a switch of her sturdy hip. I saw how she must have seemed to him growing up—do you really want a sexy mom?—I saw how it was during the war with his father away for years, and lonely boy soldiers protecting the town. "But I'd lie on my bed at night," he said, "my mother was around twenty-eight by then, and I'd hear her with soldiers. That's when I discovered what to do with the feelings I had."

Philip has dark curly hair; he looks like Heathcliff and stares mourn-fully at you, so you invent sad stories to feed him of all the young English refugees in L.A. during the Second World War, so many kids on their own. Then I sit with the nieces telling them watered-down stories of L.A. in the sixties, until Philip whispers, "How are you doing now?" We shake our heads at each other. Like his wife who they call Lambsy, I'm from another town and no one knows my family. Sometimes I help with the washing up, but I never want to show my cooking because it's too off the road and you want to play things down when you're a newcomer.

Hylda asks me how the children are—and I know she means her own grandchildren, Young Stu, Philip and Susan. I give her glowing reports based upon my memory of Stuart's talks with them and my impressions of who they might be. Like complex square dance partners, we start in one family, two figures break away, link arms, spin off, form another set, make two more, which break away, and you have all these whirling groups doing the same dance, but related in no other way.

When a driver applies for a license to drive a black London taxicab, he has to have passed "The Knowledge," the rigorous course of learning London's streets and alleys, squares and mews, which one is a dead end, which one changes names midway, and so forth.

It's no less complicated to come into a family.

We are adaptable creatures, the way we can, if uprooted, slide into a new family. After a time, we pick up the ways, the games, and customs, and even catch bits of stories to toss back and forth. A quick memory helps you mime the ways; helps you catch the Knowledge.

All the men go to the pub, except Stuart. He walks with them a while, then comes back. "So," he says to his mom, "where's your bank book?" He sends her money every month. "I want to see how you're doing."

She's sitting in her chair, by the radiator set into the empty fireplace. You think Mrs. Tiggywinkle, but the expression is Margaret Thatcher, which is how Hylda got through the death of one child and raising five others while she was working in a factory, surviving TB.

"Not on your life." She pulls herself up straight and clasps her purse to her breast. The purse all English women carry over the forearm—from the Queen Mother, who will wear it in the grave, to the Queen herself. (How do you tell the American woman? She has a shoulder strap.) Even the Prada backpack you carry on the forearm. Hylda clutches the bag tightly, "You can't look at my bank book." She looks up at him, "Only my son, Stuart, can see my bank book."

"But I am Stuart."

"I know our Stuart," she insists, "and there he is," and she points a finger at the photo of Stuart on the mantelpiece; the young man in his RAF uniform. "There's my Stuart," she says proudly, and looks her son over, "and that's not you."

"That's true," he says, "and neither is this. I asked my wife the other day and she didn't know who I was either."

"You see," his mother looks me over, "I'm right. Stuart isn't married. And certainly not to you." Just to make that really clear.

When we get back to Scarborough, Stuart and I walk on the trail down to the sea, where I sit by the starfish pool with a couple of kids my grandchildren Justin's and Phoebe's ages—I think. I look at their hands and their cheeks and remember the song lyric, "If you're not with the one you love, you love the one you're with."

But it's not true. Family is family, and you feel the difference. I don't know how. It may not be what you'd call memory. It goes back further than anything so conscious.

"Strange family, isn't mine?" he says, by way of comforting me for missing my own family, which he understands and can't do much about right now. "Or is that a redundancy?" We pick up toffee apples and he sits with me for a moment on a rock, looking out over the fishing boats coming in. "My family only reconfirms my gut feeling that families are artificial constructions—they are what you make them. I suppose I've chosen to make very little of mine, so I have very few discernible feelings."

"I see your feelings very clearly. You just don't want to go on about family like I do. To you it's a built-in responsibility. The feelings don't need exploration any more than how you feel about me. You commit yourself to each attachment in a clear, decisive way. And that's forever. It's a balanced, honest judgment of a role." I'm crying now. "I'm not like that at all." I stand facing him on the edge of the cliff, the wind sweeping my hair back off my forehead. I can't hide a lie. "Stuart, I can't give enough, say enough, be there enough. All I do falls short of what I want to give—but 'whither thou goest' I've done."

I grip his hands. "How hard it is for you with me forgetting you on one side and your mom on the other. She forgets chests of the chocolates you've sent, the great hams, the Yorkshire puddings, the strawberry cream cakes, the poems and pearls and all the letters and the love—and even you. I'm sorry, sorry for what's happened, but I'm sorry for my own as well."

We walk along the beach to Alan Ayckbourn's townhouse and then back past his theater to the hotel rose garden which is built on a cliff no firmer than the ones along the Pacific Palisades, a cliff, Stuart tells me as he holds me, "may not be here next year."

And, I think, neither may I.

18

ike any limb you've broken, you work it carefully, bring it back
slowly, moving as if you're made of glass. Like exercise, I'll do more
each day. It sounds simple.

I start by writing one postcard each day to a friend in America. It
takes forever to decide which postcard is perfect for which person. And
then what I say isn't perfect. "Hello, how are you?" isn't witty. But it's what
I see written on the card. "Just thought I'd write . . ."

A helpful friend I don't remember calls to tell me about a girl who
had epilepsy and was operated on and now has the two sides of her brain
each making its own decision. "And they rarely agree, isn't that interesting?" she says. "I know that isn't what happened to you, but I thought
you'd want to know about it."

"Wouldn't have missed it," I answer.

"I thought you'd find that helpful," she says.

"Ever so." If I knew who she was, I would send her the old postcard
from Savannah that says, "I've been meaning to write . . ."

Each time I write a sentence, the other side of my mind, the aspect
of myself that says, "You never write well in that shirt," says it has a better idea. "Go change again," it says.

I go to my writing table. I fall asleep after every paragraph. Stuart says maybe I should go to write in a place where it is not advisable to sleep.

The British Museum's Reading Room is like an opera house, filled with carved galleries and rows of bookshelves. It's centered with row after row of long tables where you sit up and read the books.

"This," he says, "is where you'll write."

"Don't think of anything else," he says that night as he wraps himself around me, "just write for yourself. Read and write."

I turn back to the L.A. of long ago. This seems easy in England, which shares the same strong class system. The accents remind me of governesses; the galleries of noblemen are like stars' houses with their favorite portraits; the parks are like great Bel Air gardens with an invisible gardener always in attendance. I can't find the finishing fever without deadline tension.

Then one day a woman I've been watching at the library comes over and sits across from me. She comes in every morning, sits in the same place; she isn't so much reading as studying, writing, crossing out, going over words, rewriting. In America after a week we would have had lunch together. Here, after a month, we are nodding. I know by looking at her books that she is Russian. She can see, because we reveal ourselves in a hundred ways, that I am an American. I want her to know that my ancestors were also Russian, so we have something in common. But my families were Jews and chased out of Russia by pogroms of her family's making.

There's an interesting difference between a refugee and an expatriate. A refugee makes me think of desperation and a brave attachment to a world that may not want you, certainly not on any terms you understand. An expatriate's an escape artist, something of a hustler, running from something. If it had been going so well, why would you have left? I think of battered linen suits and endless talks, and tables cluttered with bottles and glasses—even if the bottles are only Diet Coke and San Pellegrino fizzy water.

It has been six months.

Her name is Svetlana. "So," she says one day, "I know you are a writer." She has astonishing long fingers, like bony serpents; she needs to be a star.

"No, I was," I say, "I'm just practicing now."

"I don't think so. And I want you to show me. I want to tell my mother's story. This is important."

"I can't, really." I don't want to tell her we can barely talk, and I can't remember writing—can't hold onto a visual image—writing's even harder. "You should do this with someone who knows you both best. Your best friend."

She touches her chest, and with a sweep of her hand spreads her fingers across my breastbone, "But this is you—you are my best friend."

My mother's maiden name was Svet, so Svetlana is unforgettable. Svetlana writes in English with a strong Russian accent, which is hard to pull off. She has a harder time finding words when she speaks. She smacks her lips, furrows the brow. I'm with her when she loses the word, but I never dare to come up with it; a taboo thing, throwing in someone's missing word. "It's good to let me try, the hard way is how to remember," she says. "Invitation!" she exclaims, "come to our house. I will tell you where it is and you will write it down."

The Englishman is appalled. "You know she's Russian? This is her address, her husband's name is Sergei, and you met her at the Reading Room? These are all excellent reasons for going over to a stranger's house for dinner in a neighborhood I don't go to." The English accent was made for dismissal.

"She's in the theater, works with top artists, and I met her at the Reading Room."

"It's not a café, you met her *in* the Reading Room, and just because someone reads and speaks two languages, you assume she is smart and would make a perfect friend for us. Not for me."

I explain Laurie is also invited. "She will meet people."

"Good," he says.

But he goes, and on the way, Laurie asks what we think about Jennifer, whom she'd brought over for tea. "It's easy to see you're in love," I say.

"But she's clear," Laurie sighs, "she'll never bring me home to meet her parents. I'm Jewish, American, and gay."

"And an artist," Stuart says. He prefers to stay out of these talks when he knows nothing he says will change anything. But he adds, "You'll never keep her in the style she's used to."

"But she's a banker. She could keep me!"

"She won't. She's already putting up limits," Stuart says. "I'd say just have a good time and don't expect her to change, but that's not how it goes with you."

"You're projecting," Laurie says.

"On experience," he says. "That's different."

"Not seeing her anymore is hardly an appealing idea," I say. "Something will work out. What is happening with the new gallery?"

"Not that you're changing the subject," Laurie says. "He's coming back next week with his partner."

"This could be exciting," I say, "falling in love with a banker, and a show, all at once."

"You have lost all sense of reality," she says. "You have to keep her inside," she tells Stuart, "she can only do damage."

Stuart is sitting on a white garden chair next to Svetlana's ninety-year-old mother, shadowed by one of the dark lilac trees in Sergei and Svetlana's garden in Clapham. There are soft potato pancakes, borscht, fish cakes with dill, dumplings, and cucumbers for dinner. Svetlana's mother has my mother's long slender jawline that grows ever more arched to compensate for the posture she can no longer adjust. She is alone during the party, looking grim and desolate until Stuart sits down beside her.

He's talking in the international language of gestures. He doesn't put on acts to the extent I do (hard to do), but this—Poet Revealed—is one of his best. Just the way he was when he kissed Ninette de Valois. He introduced me to her after the ballet one night. He adores elegant older women. "I had a book of hers, the *Penguin Book of Ballet,* and I fell in love with the book." Suddenly, he's reciting Verlaine and quoting *"Dans le vieux parc . . ."* in one sentence, and having things to say about it, too. They patch together a conversation in French. Svetlana's mother is now sitting up straight, looking him up and down. I remember this expression of his. I remember him meeting my mother and the same thing happened. He is good at courting mothers.

But I don't remember how he courted me. I don't remember how we met. The deeply blotted-out patches don't seem to have a form. It's as if a bottle of ink was spilled from far above; and very occasionally, a bottle of remover is spilled, bringing something else clear again.

After supper, friends of Svetlana's play Russian songs under the willow tree with the enthralled expression real musicians get when they play. I have never seen a writer smiling at work. I am watching Svetlana's friends and, like the Cheshire Cat in the tree, I remember Michael Tilson Thomas playing the piano with this same look at Corinne's on a Friday night. Stephen Sondheim also looked like this the summer he visited with his mother. Stephen smiled when he played. Not when we took him to the beach club, a new club, the first one Jews could go to. He went for manners, a big thing then. And then my cousin Julius Katchen used to get this glow on his face when he'd come out from Newark to be with us for his holidays. He'd practice, smiling, on my mother's Steinway. He was a prodigy and Mr. Chenoweth, our piano teacher, was breathless with admiration, and Julius would go across the street to visit Artur Rubinstein for tea and recitals. Artur's daughter Eva would greet Julius at the door, towering over him and looking at me on my bike like I was the chauffeur. You'd think Julius could not possibly eat more, but they'd give him cakes there, which he'd tell us about after he came back home with this same smile.

"I have a cousin who's a concert pianist," I told Michael when we met, "you'd like each other, I think." They would hate each other, of course. How pleased I would definitely not be if someone came up to me and said, "My cousin is a writer and you really ought to meet."

Michael tried to be pleasant. "A wonderful pianist," he said, "Brahms especially."

"So then you know him?"

"Yes," I said, "and Julius has a son who is, I guess, really troubled."

"I can think of a reason right why he might be troubled," Michael said. "Julius is dead."

Stuart slipped in here and gently reminded me about our trip to Paris to visit Julius's widow, a tough Parisian who had been living with an Algerian man.

That night I remembered only the widow's legs, bare, tanned ankles in old high-heeled shoes. Tonight I remember all of this as we sit in Svetlana's garden, clouds the shade of creamy tea drifting by behind the poplar trees, black silhouettes against the sky still larkspur blue at midnight.

"You are so low and quiet," Svetlana says to Stuart. Svetlana's one of those women men who don't usually talk will talk to. European women see men's humanity more than I do. If a man caricatured me in the way I tend to do to the men I've claimed to love, I'd dismiss him as a chauvinist.

"Odd days, that's all it is," Stuart says. "I'm waiting for something to break. And Jill's wrestling exhausts me," he adds, and looks at me. "I know it's hard for you, but sometimes it's even harder for me to watch. I wonder, can you handle this? Should you? Maybe this suspense is the feeling you have before dying. Waiting for Godot. Or maybe it's all the deadlines."

"You are both doing this same thing," Svetlana says.

"No, he's been cheering me and everyone else," I say, "his mother, his children. He's planning our next trip and living for next year, the next paycheck."

"Maybe," Svetlana says, "you need to both stop." She looks at us closely. "You are adjusting all the time. No wonder he's exhausted."

I don't wait for him to answer. "So maybe I have to be more independent, whether he wants me to or not." I turn to him. "If I just go ahead, you'll see it's all right."

19

I'm on my knees in this garden in England's long twilight, a choir singing inside the church, and I'm praying to believe this makes any sense. A happy voice filled with American confidence has invited me to come to High Table at Oxford.

This probably began when I got an invitation from Ron Clifton, the cultural attaché at the American Embassy, asking me to serve on the Fulbright Commission and to be in charge of selecting the writer for the Raymond Chandler Award.

To receive a Fulbright was very impressive when I was a teenager. The grant funded you so you could stay in Europe and work on a particular project. Senator Fulbright set up the scholarships because he'd been in Europe during the war. He'd realized that few Americans had a sense of how it feels to live in the rest of the world. We like to torture ourselves with this, to add it to the list of things we hate about being Americans so we don't get too grandiose, but not a lot of Slovenians, Finnish, Chinese or French like anywhere else either. If you can live somewhere else, it gives you a perspective, an edge, and a new tolerance for irritation, which is useful.

To be a Fulbright Commissioner was even more of a very big deal.

For four years you'd help choose the awards and go to events at the embassy.

"Do they know I didn't finish Stanford?" I asked Stuart.

"You've written five books and hundreds of . . ." He doesn't need this discussion. He's reading.

". . . of articles. I don't want to hear the litany of stuff I've done again. That's not the point."

"That's exactly the point," he says. "You're bored by that, which means you remember it. Boredom is an excellent sign." He looks at me over his glasses, the right brow gathered up in circumflexes. Where have I seen this expression? Yes. Victoria. Am I always irritating when I'm writing? Even when I'm actually not?

And another problem here. How do I choose the best manuscript for Fullbright when by the time I read one sentence, I have forgotten the one I read last? That's the short-term memory problem. When it's this bad, you don't have what they call a working memory, one that can keep more than one bit of information on hand.

Mine is unemployed.

So give it a job, and it will figure out how to do it.

Handing out a grant for the best detective novel proposal couldn't be better.

I'm looking for the memory. The writers are looking for a university where they can work in exchange for giving a few lectures.

All of us are detectives. However, the mind has its own shorthand. When you're reworking a memory, it needs coaching. This is a perfect project. As I go through proposals of plots, I draw images of characters. Which one comes easier? Who is faster to cast? What dialogue grips me? The project isn't so automatic for me. I bring no preconception to meeting these characters.

The wonderful thing is that certain events are fresh each time. Oxford High Table is one of those. Larry Beinhart, the second writer I chose for the Raymond Chandler Fulbright Award, invited me the first time. He came to lunch at my house last Sunday with his kids, so his wife, Gillian, could work on her writing in Oxford. Noah showed his kids the walls to paint on.

Everyone at my Sunday lunch watched Larry carefully, the way we'd watch new visitors to John Lahr's teas on Primrose Hill. Every gathering you go to has an element of suspense. Will this person show up with that person we all hate? Will he really bring him when he is there? Will she tell that story again, and in front of her? The more of us who were there and the more restless we were to get on with our own riffs, the longer everyone else's chat would go on and the more difficult questions about increasingly obscure political issues would come up.

The elements of suspense at our Sunday gatherings are all about our writing. The routine now is to bring three pages of the same project consecutive to what we were writing the week before. We read aloud to catch if the rhythm of our writing feels true to the pace and tone of our voices. My writing feels like cold cream labels. Does the first page of what I am writing have any connection to the next? Will Geraldine have a rock concert drug scene, a sex scene, or both, and will Mark have a new subplot set in a different Asian embassy this week? Will the sharp new writer Bramley be writing a screenplay or a novel? And will I remember what anyone read last week?

I can't remember who has invited me to High Table this time, but unlike the situation with the person who was also Erica, I can guess it has something to do with Fulbright.

Since I'm wearing my old Ralph Lauren commander's jacket, several Americans at the Paddington train station asked me what time the train was leaving and where is the track for Liverpool. I might just as well have told them, because people I ask say the time has changed or they're not familiar with that route. Americans like to go to Paddington. They think it will be a wise little bear of a station. It's a huge, up-to-date station, complete with a flashy overhead timetable crawl and a cappuccino bar—which in no way affects its authentic eighteenth-century operation.

After maybe forty minutes on the train to Oxford, looking out over the rolling hills and green trees, I think I am in northern California, arriving at Stanford University with my parents in one car and my suitcases in another. I'd been accepted solely on the grace of several short stories. I had no real school background. I'd been tutored at home during the war. I thought it was because of the war. I remember staying overnight with my friend Judy Lewis, Loretta Young's daughter. We went to a Catholic

Church for Mass the next morning. The priest rang a bell and I thought it was an alarm that there was a Jew in church. The next thing I knew, I was in a hospital.

The same thing had happened at Chadwick boarding school. I went out horseback riding and woke up in a hospital. I was told I'd fallen off the horse and had a mild concussion. I remembered riding, moving from a trot into the rocking force of confident canter. I didn't remember much after that, except leaving school (which I didn't mind) and being set up with a program of tutors from my old school who had organized a plan of classes that would, exams permitting, be acceptable to Stanford.

Once there, I couldn't memorize by rote, which made science and math classes difficult. I see now I memorized history by visualizing characters that would disappear as soon as the exam or the paper was finished, but my writing grew in assurance. One day I was out driving, faded out, drove into a truck, and once again found myself in a hospital, once again on phenobarbital, which made me drowsy. I could not stay with the stories my teacher, Hannah Greene, told me were so full of promise. This was a reason I was given amphetamines during the day.

All of these instances of passing out were rationalized, excused and no connection among them was ever made. This was only natural in the time I grew up, when epilepsy was seen as an unmanageable and unmentionable disease, like alcoholism.

So even a visit to Oxford feels like a very big deal, in case I have not reminded you for a moment that I am here, walking up to the gatehouse entry at Corpus Christi College. I'm forty-five minutes early. The gatekeeper says I can walk in the garden. On the way I walk by a chapel and hear singing. I slip inside, and listen to students rehearsing hymns. During a break, I walk along an arcade winding on. I wound my way through lavender, forget-me-nots, sweet alyssum; drifts of mint and thyme softened walls of stone, along with pansies, pinks, and deep velvet sage. Safe here, with the privacy I need to pray, I kneel down and ask whatever force is listening to help me come up with words fast, to be buoyant and fresh, not to sound like an idiot tonight. Prayer works for setting me firm on a course of concentration and helps make me less self-absorbed.

There's a tap on my shoulder and I see a tall, gaunt man, his long black cloak drifting behind him, falling open over a white shirt, a tie, and over that, like a bright weskit, an Andy Warhol t-shirt. "Your friends are waiting for you," he whispers. I follow him carefully to the gatehouse and, with the bravado it takes to be an American in Oxford, I switch on a big smile and meet the Hallorans. I launch into my memory riff, the fast version.

Susan reassures me, "You explained about your memory to us. It's okay. John's here on a Fulbright and we met at the embassy last year."

It was too appealing to be a Fulbright Commissioner, granting the Raymond Chandler award, to point out that I wouldn't be able to remember an applicant's book from one page to the next. I would have to figure out a way to do it.

By making notes, as I read, on what I thought was fresh and interesting—even surprising—and adding up how many times during the reading of a book I stopped to mark that down, I was able to choose the winner. The one who gripped me the most times won.

John Halloran wasn't a writer. As we walk to the dining hall, he explains in what is probably his short riff that he's a scientist. "I'm working on a project where we'll be able to take an object, say this glass"—we've come into the huge, medieval dining hall—"describe it to a computer, which then speaks to a computer, say in L.A., which then creates an actual replica of the glass there in L.A."

"John's particularly happy," Susan says, "because he's been working with sculptors, and the latest thing," they beamed at each other, "is you can take a bone that's been crushed, a jawbone for example, and create a new one to replace it."

I turn to the man on my other side, James Griffin, and ask what he does. "I'm a philosopher," he says with sort of a challenge in his voice. I looked him over. He's said it maybe five thousand times in his long life and has seen this blank look, likely followed by "Oh," and a quick switch to the weather.

"What's your philosophy? Or do you teach all of them?" I ask.

"I'm a philosopher of values."

"That's difficult right now, to find any to think about," I say. "Is it likely that since my values have changed since I was in my thirties, I will shift my memories to suit my values?"

"Possibly," he says, "and you may not be aware you're making that shift, since you may not remember what your values were."

I stop eating the comforting roast lamb, served this way since the beginning of time, and look at James. "You're an American."

"Yes," he says, "they're fond of us here, if we keep a low profile. I've written forty books and the university's published all of them. I've been here a very long time."

I've been watching the new writer, Anatole's daughter, Bliss Broyard, all evening. I can't believe she doesn't know me, but perhaps Sandy, her mother, has told her about the memory and she may not want to embarrass me.

We move to another room for dessert, where James says, "If you like, I'll introduce you to the young woman you have been watching all evening."

It's not Bliss. It's Andrea Ashworth, whose stark violent memoir, *Once in the House on Fire,* has just been published. A friend of mine, Laura Roman, a junior fellow at Oxford, said, "You have to. Yes."

Andrea's complex story about her parents, her battle to get away to Oxford grabs me. "My memory holds when it sees what it wants," I tell Andrea, "and what I want is to read more, to catch that force you have."

I don't want to leave Oxford. "I think if someone gave me a course to teach and a room with a bed and a desk, I'd stay," I say as we move for the dinner's third act to the sitting room for coffee.

"What would you teach?" James Griffin asks.

"Hollywood myths and legends—the fictional history of America as seen through its filmmakers' eyes. Something like that," I answer.

That's not what I'd teach at all, I think, standing now in the Oxford station just before midnight surrounded by loud, crazy students bashing each other around. This isn't the Oxford you have in mind, I tell myself as I duck to miss a flying bottle. I'd teach the art of memory. That's what I'd teach.

I think of John Halloran's great new magic trick. I can do that. I can take this evening, put it in my particular mind and it will disappear. Voilà! Or come out in such a way that no one who was there will recognize it at all.

James Griffin, the philosopher I sat next to at the Oxford High Table dinner, wrote me a note, suggesting that I visit Dr. Martin Davies, a philosopher with an office in one of the psychology departments at Oxford.

When the train stops in Oxford, I don't look at the note I'd made myself because I know I remember where I'm going. So I take a taxi to the wrong college, where a girl from Texas who went to Yale tells me where I should be—not Christ Church this time, but Corpus Christi College, one of the smallest colleges in Oxford.

I follow Dr. Davies through a warren of crannies and up several flights of steps to an office with a small table for six and a few hard chairs and one easy velvet chair.

"James Griffin has the impression I work in memory," Dr. Davies begins, "because I have an interest in acquired language." I wonder if acquired language is one you buy when you forget your own. I keep this confusion to myself. He continues, "A philosopher such as James has an interest in perception, which is how we learn about the world now. Memory, of course, is how we learn about the past."

Martin Davies has a lot of thick, dark curly hair that he keeps lacquered down, a mustache, sharp eyes, and the eagerness of the ten-year-old just discovering his genius, the ten-year-old who keeps his delight at it.

He presents each comment like an unsettled idea he's heard something vague about. Then, as you listen, you pick up that it's a major new theory he's learned everything about. "To begin with," he explains, "there are two kinds of memory. The first is semantic, or factual, where you know the meaning of things without any memory of how you learned it; as in 'a rose should smell like this.' The second is episodic, as in, 'I became uneasy with dogs when my mother's dog bit me when I was four years old.' You learn because of a specific event."

"For example," I say, "I love this garden because I was in my new bare sundress, sitting up in this olive tree, looking out over the pool, when Clark Gable came through the gate to see my father? Is that what you'd call episodic?"

"Colorful episodic, a perfect example," he says. "It's also called a memory of particulars."

"So, I may be learning something here. A semantic memory might be that I love sitting in olive trees. Or am I confused?"

"No, your memory of association may be gone, but your aesthetic—or neurological—memory is there," he says. "You'll remember, say, that olive trees have a sultry charm to them, a protective, romantic glamour, and I like sitting here."

"So we're both fond of sitting in olive trees," I smile. And, now, as I write this, I remember it wasn't me seeing Clark Gable, but my friend Judy Lewis. This was the story she told me. The story I wrote about in *Perdido*.

"If someone has no memory of dates or events, such a person would have difficulty having any concept of a past." He looks me over.

"I may be almost such a person," I say. "Would it help to restore my memory if I were smarter, or are there exercises, things I can learn?"

"I wouldn't confuse memory with intelligence," he says, "although I suppose if you have a high IQ, you'll devise new skills to hold and retrieve information—if, that is, your skills are okay. You can refer to skills, or routine tasks, as hard-disk memory, although I prefer the term 'procedures' instead of 'skill.' Sometimes we don't even know we keep them. You might be asked, 'Can you type?' and you'll say 'No,' but put a typewriter here and you'll be able to do just fine. Sometimes short-term memories are important enough to put into the stacks."

"Like in the Bodleian Library?"

"Exactly. Some people remember everything so firmly that there isn't room for a new idea. There are people," he likes this idea, "who speculate that because autistics don't always see how things work together, they're able to break out of their mindsets more freely."

"You know, I couldn't recognize my husband at first. Now, sometimes when I see him after a couple of hours away—like when I come home from Oxford this evening—I'll be startled by him. Is that odd?"

"No," he flutters his hand, dismissing the seriousness of the condition, "you at least see the problem. One patient was so bad, I'd have her take the tests, she'd fail miserably," he's laughing now, "forget how badly she'd done, and talk about how good her memory was!

"Here," he says, "try this. Take three letters of the alphabet and hold them in your mind."

"Okay, S, Q, and R."

"Now, count backward from thirty."

It works. I'm pleased. I've remembered the letters.

"Like any exercise," he says, "it doesn't seem to have a point until you need it." He's quiet for a minute, going through files of ideas in his mind, like the piles of books and papers scattered around the room. Then he continues.

"There's a book about face recognition by Vicky Bruce and Andy Young, published by Oxford University Press, that's tied in with an exhibit of faces at the Scottish Gallery. You'll find memory's information there, like in the Bodleian.

"There are two systems of recognition, the first of which is 'who is it.' Face recognition may have a separate neurological system. Even though sometimes faces go and stay gone, they may also come back."

What he's said means, of course, it goes both ways, but he didn't want me to be discouraged. "That's one system of recognition, 'who is it.'"

"Then," he says, "there's the second system of recognition, 'how do I feel about it.' You might not know it is your husband, but you know how it feels. And there's another thing," and this interests him. "It looks like your husband, but you don't feel what you expect to feel. Someone's taken your husband away and they've put this ringer here. I remember a case like that. This woman believed a robot had been put in the place of the person she loved. It became pathological. To prove it, she decapitated the person!" He's fascinated by that.

"Then there's another story: Here everyone else—the daughter, the son—knows this is not the woman's husband." Like a wizard from Yiddish folklore, he sees science as a story. But, it's fine with the woman—"We get along," she's told him. "She'll deal with it."

"John Marshall at the University of Psychology talks of situations where you remember episodes, but no longer get the buzz. For example, you know this is who you are married to, where you fell in love, and how it was, but," he shrugs, "that may be less to do with memory and more to do with electricity."

"Or you may just fall out of love," I say.

"And it has nothing to do with memory. But," he looks out the small windows overlooking the towers and trees of the campus, "the skin conduction has changed. There's something called galvanic skin response. If someone loses face recognition, you can test the skin and the electrodes will show if this is your mate."

I remember playing with the children on the beach at Scarborough. Their hands touched mine as we built sand castles, and I closed my eyes, imagining they were my own children as they dusted sand off my cheeks. But they weren't my own.

"Maybe that's almost a genetic characteristic," I say, "the 'touch' of your own genes."

I'm so interested in talking to him that my own memory's disappearance seems beside the point. So when I went over these notes, the meeting with Dr. Davies comes back quite clearly. Maybe the less stress distracts it, the more memory holds on.

Before I leave, I ask him, "Would you have any idea why I'd rather draw now than write?"

He thinks about that, listens to some of my background, and says that if memory's gone, whatever inhibited me from art would also have gone, and the art may be showing through. This leads us around to the earlier theory about autistics and originality.

About short-term memory, it's hard to find the logic. By the time I had left the room where we talked, I couldn't tell you what tie Dr. Davies was wearing, but that was a "fact," and only on the soft disk. I could, however, draw a picture of the chair he said I should sit in. I probably made a character, an image out of it, thereby putting it away in the stacks.

And I have not forgotten that Dr. Davies says, "Your memory of fear can also be gone." This explains why I'm not afraid to travel alone now—or was the fear a learned response, and never basically there?

"Not only," I say to him, "did I make this trip to Oxford, but I did it wrong and was still able to get to the right place after all." I write that down to remind myself that there's no disaster if I go to the wrong place. Go early enough. Leave room for mistakes.

The basic way we deal with a serious war—I don't remember this, but fall into it by reflex—is to move into silent retreat. I am lying firmly on the far side of the bed the way you do when you haven't made up.

Do the opposite of what you feel like. I turn to Stuart, stroke his shoulder, and look across him out our window to the full moon shining through the tree outside, the light outlining his jaw and his arm. I could frame this window, heavy lace at the edges, furled like sails with rough ropes wound into strong curving knots.

I'm back draping one of the small mannequins I dressed with fabrics my aunt, Hope Skillman, sent me from her company in New York. Her catalogues on the floor become storyboards I've put up with lists of costumes we made for Warner Leroy's Ram's Head Productions at Stanford. I dyed the white ballet tights by filling the bathtubs and lying down in them in the dye to make sure the color came out even. But I couldn't master the sewing machine in freshman-year costume design. You can't be a great designer without sewing perfectly. You can't cut offbeat if you can't cut a straight suit, can't do a riff if you can't write a declarative sentence. I turn over to him and touch his shoulder. Even in his sleep, Stuart always turns and puts his arm around me.

He must miss his jazz friends. So many are gone. Gerry—was it just last year? We must visit Franca in Milan. Stuart fits around me as easily as an Armani. Can you play jazz if you don't know your scales?

Aunt Hope wanted me to be a designer, my mother wanted me to be an artist, and then I watched Barbara Bel Geddes in my father's picture *I Remember Mama*. The camera froze in closeup—showing memory at work as Barbara told how she was writing up the memories of her childhood and her mother—then went off focus, dissolving into the next shot to show you the past. A perfect flashback.

Try this. Go to a mirror. Stare into your own eyes. Focus on what you can catch of the scene you want, even a corner. What you are wearing. Move in even closer. Layer today over yesterday, slide today away, so slowly, again.

Cut to black. Try again. Like striking matches on jeans, the moment you want doesn't light up at first. You have to believe the whole scene is back there on file.

Barbara Bel Geddes was older than I was, but not so much that it was one of those adult ages I didn't ever expect to be. I wanted to be East Coast elegance, like Kitty Hart in black velvet. Or to be a tough woman from a Mitchum movie with a straight whiskey voice, like Jan Sterling. I'm leaning forward over him. Was she married to Paul Douglas? Edie Wasserman had powder-white matte breasts, smooth above the draped bodice of the black crepe cocktail dresses she wore with diamond Tiffany clips, like quotation marks, on either side of the bodice.

Some had breasts shining with suntan oil, like Sidney Sheldon's wife. Sidney had a stud look. You could see him doing real estate deals. He was one of my father's protégés, but he wouldn't speak to me when I was trying to write a biography of my father for Michael Korda, the ideal editor for that one. He said my books had only been an embarrassment to my family.

Killer cuts have mnemonic overdrive.

God talked to me in this dream and made me fly and said writing's not a lot different from breathing. Neither is this. But how does God feel if you're thinking about Him while you're doing this? God invented the idea. Loving—writing—all—like breathing—works as long as you don't for a second believe you can't do it. Hang on like it's a glider. Do you think you can't go here or there? Slower now. It'll go as long as I stay in close to the heat—to the power source. Keep going, long, slow, and steady. Am I thinking of loving or writing? You can do it, are doing it, it is a song, and another one. I go and go, up over everything, never stop. Flying out, out over the coast.

I do actually love this person on automatic, like I love my father, and my room, and my children. I stop. Chill out. Like they used to say on juvenile delinquent warning signs, DO YOU KNOW WHERE YOUR CHILDREN ARE NOW? DO YOU KNOW WHO THEY ARE?

I pull away. "Are you okay?" He touches my hair. "Fine."

You can't be there and be here. Don't do that. Don't think of home; this never works when you have any of the kids on your mind, mainly yours. But deep as a sword's slash comes the longing to see Phoebe's face. I want to catch every darker new angle to my grandson Justin's voice. Did he get the cricket uniform I sent?

Where do you think the big romantic stories, happy endings, come from—you fall for your parent, you come so close you singe the feathers. I love you so, I'm thinking, but I hate the feeling. He'll never get off unless I think of models with endless legs, how do they say it; limitless, legs through the roof, over the top, did he say Picasso lived so long because he did this every day, or did Picasso say it? It doesn't matter. Picasso isn't here to say that he never said it and Paloma wasn't there to say yes or no; and it doesn't matter yes, because he's believing it and that's the wonder, when I think of it. He's here.

Look to this moment. It's getting light outside.

Look to this day. When I was in school, we used to stand on the lawn with our arms raised to the sun and say, "Look to this day, for it is life, the very life of life," and the day I'm having will give me the way to find the memories I need. I can write as long as I like, like the songs in a big show, the words will keep coming and coming. I'll have everything to say, everywhere to go. Don't ask how you get here or if there's something beyond beyond—because it's there. He has it.

22

*D*oes personality remember who to be, how to be, if you've lost everything else? Or does it watch what it sees and find new riffs, a new voice? *Listen,* I remind myself, *the best way to find what you need is to construct a strong new present.*

I don't want Lynn, my agent, to know how bad it is. Angie works with me, walking me through chapters she maps out the way you'd take a blind dog through a park.

I haven't told Lynn Nesbit that French *Vogue*'s editor, Joan Juliet Buck, called from Paris last Thursday and asked me to do a story on Liz Hurley.

"She was really friendly," I told Valerie Wade, one of my best friends here, by which I mean she does not expect me to remember what she tells me or that I have been over to her house. "That," Valerie says, "is because Joan and you have known each other for years. I think you met her in Leo Lerman's office in New York."

"Great," I said to Joan. Presence of mind here. I did not ask who Liz Hurley was. Or if I did, I do not remember what she said.

Lynn knows. She called Friday. "You're throwing away your work on little magazine stories." This is exactly what I need to hear. More, more. She has a closet of tough lines tailored to each of us, like an S & M couturier in Soho. I bet I have a tough line from her, like a caption, under every piece of my work. She's the teacher you really learn from. Others threw me out of class, but she's sticking with me.

The writers and Judith are here this Sunday. I am spending more time on lunch than writing. Today it's chicken paprika and poppy seed noodles.

"You're cooking, not writing," Geraldine says.

"I try to tell her that, too," Stuart says. He has come by for a plateful.

"This is not stopping you eating," I point out.

"Give him a break," Judith snaps. She is his person. If we are surrogate parents, I am the eldest daughter's annoying mom.

Judith has brought Stilton cheese and grapes, which Stuart loves. She goes in the refrigerator, takes out the terra cotta dish with foil. "Polenta? Do I like that?"

"You could try it. I'll heat it up."

"And some of the, what is that?"

"A peach crumble."

"Yes, with a little ice cream. I've got to stop eating."

"I think I'm leaving," Stuart says. He returns to his dark gray room where he'll spend the rest of the afternoon.

"The thing about him that is so remarkable," Angie says, "is he's here even when he isn't. His personality stays in the room after he's left."

"It's true," Geraldine says, "he's the center of the room even when he doesn't talk. His is a stalwart presence."

I must discuss presence with this Professor Steven Rose—if he answers my letter. But then, presence is not the part of the brain he studies. He is the biochemist who has written *The Making of Memory*. Memory and presence. Is presence a brain thing or a body thing? Or is it spirit? I think of Stuart when I should be thinking of the work. Does he matter more than my work? That might be new. If I don't remember my old character traits, can I figure they're dusted? And I put you first. I really do love.

Stuart's particularly grim today because Geraldine has her kid, Sam, here and Noah is upstairs painting.

"What does Liz Hurley do?" I ask after we're settled around the table.

"She's a model, gown held together by safety pins, and she's having an affair with an actor," Geraldine says.

"By the time you get the story done," Judith points out, "the rest of the world won't remember who she is either."

"You just say that because you can't wear the clothes she does," Mark says.

"Or doesn't," Geraldine says.

"That's not your kind of thing to say," Mark tells her.

"I'm testing a new character," Geraldine says.

I can't remember whether Geraldine's new character, Deodora, is the star or the manager in love with the star. Is that the writer's problem or mine?

"If French *Vogue* is doing a piece," I say, "then she must be someone. But I will be doing it for the wrong reason. I want to be translated into French."

"You are already affecting a French accent," says Judith, "and it's bad."

"But, of course."

"You could ask her how she'll feel being remembered for one dress."

I won't. I show them the picture of Liz Hurley they've sent over. She does not want to be remembered for the curve of safety pins coursing down her body in Versace's dress. I feel the tracks of those pins, the quiver of tension in the taut pull of the dress. The curve is like the snake draped around Eve, outlining to Adam what he's not likely to get (unless he really toes the line).

"I'm not giving Versace, any Klimt or Erté title so fast, but this dress on this woman is about now."

"But will you remember 'now'?" Debbie wonders.

"This has taken you away from your book," Geraldine says. Geraldine is worried about the book in the way I forget to be.

"The book's going through a personality change."

"We know that," Svetlana says. "How do you say, 'in retreat'?"

"True." It is, but I don't want her telling me. It's okay if Judith said that. I should try to remember that when Svetlana's sharp it means she's needy.

As if to prove what I mean, Judith remarks, "You have a new habit—I'd watch it. You look in the mirror all the time. You have a lot of them around and now you don't miss a glance."

"Thanks," I said, "that's reassuring." I look at Noah's new drawing of another rocket diving into the painted nasturtiums on the top landing.

Judith has had highlights put in her heavy, dark curly hair. The highlights give her eyes glances of gold I can't figure, as if the light hit the expression center inside. It's very sharp. Just when I decide I'm intimidated, I always glance at her delicate lower lip, trembling slightly—will she be in control no matter what comes up? What you say may hurt. Worse, what she says, before she realizes it, may hurt you, and she hates that.

Someone has given me an architect's book, *Chambers for a Memory Palace*. Maybe with memory coming all around me, with learning about it, I will lure it back. Like an act, if I understand how it works, what some ideas are, how it looks, I will at least give the impression I have it. If I can remember, that is, what I am learning.

Memory palaces are a variation of the memory theaters I first read about in Steven Rose's book. The ancient thinkers hated writing and worked out many ways to have their ideas and stories remembered forever without writing things down (for themselves, they had scribes).

Plato said Socrates thought writing was inhuman and established outside the mind what could only be in the mind. And Cicero thought writing weakened the mind. "You put it down, you lose it."

By the way, the mind, Steven Rose makes clear, is not to be confused with the brain and its properties, structures, processes. The mind has sensations, thoughts, emotions. "The mind is the dramatist, the poet. The brain does the homework, edits and remembers structures."

Yes, Gavin, the architect who lives downstairs, brought me the book, *Chambers for a Memory Place*. Donlyn Lyndon (with Charles Moore, who died before the book was finished) writes that two thousand years ago, before Cicero would make one of his two-hour speeches to the Senate without notes, he'd construct a theater in his mind filled with balconies, rooms and arcades. He'd go through it in his imagination, setting around furnishings and statues to remind him of each specific idea he was going to talk about.

Ideas and events, allied with particular images and their distance, their placement and proportion, became memorable in spaces that had clear, recognizable, permanent order.

When I read about this ancient understanding of memory, I was reminded of *Fahrenheit 451*, Ray Bradbury's story about the loss of literature. I saw his characters in togas and sandals wandering through arcades, making sure the stories, poems, and novels are remembered. The arcades are draped with ivy and wisteria; images representing the characters, the ideas, and the values are set into sun-dappled niches, or up on shadowed balconies like those around the Bel Air Hotel.

I want to do that with my years; give a year to a friend and be able to call up and, "Okay, February 1967—so where was I?" I'd like to have it clear what we thought that year.

The ultimate book Steven Rose recommends, *The Art of Memory* by Frances Yates, is the exquisite, detailed history of all the ancient techniques of using your imagination to place ideas and images in order, all the medieval and Renaissance approaches, the masters' names and techniques. I am inspired.

I will put good lines into niches, characters into balconies, scenes on stages, and garland chapters along arcades, so I can simply walk through my place and pick up what I want to recall.

While everyone's quietly eating, looking at the memory book, and going over pages, waiting to see who decides to read first, I look around my rosy red room here on Wimpole Street with red and magenta candles lit, the fragrances of dark pink lilies, sage, the onions, the paprika, oranges. I don't know where it's all from or how exactly it was put together, but what I'm not sure of yet is, will this remain even when I'm away? And will I remember it when I am not here? Will I be able, like Rose explains, to store what I want here on this shelf, or in this silver tea caddy? Little by little, will I put more things and people I need to remember around my house? As I can handle it or as it becomes necessary, I will, I suppose, put up shelves, bring in extra galleries and invent an attic with chests and old statues representing scenes and moments.

To begin, I pasted up family pictures around the walls in the kitchen and my workroom. This has become my memory theater, and as the days go on, I can tell you before I come into a room whose picture is where. I am not certain where each picture was taken, but in no picture is it raining.

Stuart is the lead player. Most of what I do remember about him is what I need to know, which is that he is always here. I have no fear that if he goes out he won't come back. But if this is true, why does my breath catch each time I see him? The parts of my life that are almost completely blank are these recent years with him. Then the years with my children when they were on their own with me are faded, shadowed out. Negatives held up to a hot sun. I have them in my theater on the stage, to the right and to the left, giving them independent entrances and exits. I must stop referring to them as "the children." This gives each child his own spotlight, an accessory, an image. Johanna's smile, her rosy cheeks. Her husband, Rob, the fireman, brave, silent. Justin, my grandson who loves baseball and hates talking on the phone, but no more so, Johanna tells me, than Jeremy, who is here with his daughter, Phoebe.

"Memory theaters," I tell him, "refresh and organize intimate family relationships."

"Intimacy and family," Stuart says, "are not to be confused."

"Especially if you don't live anywhere near your families," I remind him, "and, therefore, cannot make them intimate."

This was still the serious weekend issue. I miss home, especially childhood Sundays, filled with people playing tennis, swimming, eating deli from Nate 'n' Al's, eyeing each other and going for tops in charades—fast lines, low cuts (dresses), long looks and salty digs. (Doris Vidor, Linda and Warner Leroy's mom, always won at those.) I can't exactly place myself there. But that may have been the point. I stayed outside—watching.

Stuart says I tried to imitate those L.A. Sundays when we lived in Connecticut, and now again by having my writers' group here on Sundays.

"Possibly," I say, "so now we can fight about me missing my kids all day on Saturday."

After the writers leave, Judith goes into Stuart's room and sinks into the leather couch, holding the needlepoint cushion she made him.

"Cesario can't leave his wife or his work around my birthday so that we can go away. He is simply not going to be here for me."

Maybe Judith needs the anger. Fear's paralyzing.

"Maybe, Judith, let him go." I hedge when I'm firm. Is this my style?

I do remember, in *Bed/Time/Story*, not of course in my real life, someone telling me to dust the guy. And it was hard for me. Now I see it was hard for my friend to say that, too. We don't like to tell anyone to cut the losses when the loss is love.

I've gone too far.

Tears.

"I can't."

"So why ask me for advice?"

"Because I do. Then you tell me it will never work and that won't help."

"Sometimes all I want on a Sunday is for Stuart to say I've got a ticket to go home. You aren't the only one who feels like hell."

"Maybe," she says, "someday you'll get your book together, you'll get your own ticket for the States—now, there's a concept—and you're going to say, 'I'm going home.'"

But is that what I really want? Or just this weekend?

What Judith is actually here to talk about, until Stuart's finished going over this latest contract from Laurie's new gallery, is Cesario.

"Like all intriguing men, he wants you to be his enchanted mirror. He'll polish the mirror, give it things to reflect upon. Do you want to be a mirror?" I look at her hard. Turn it around: this way she has complete independence.

That night Stuart and I are in our separate workrooms. I am not working on the book. I am playing old movie scores and working on my memory theater: my father is on the aisle in the front (this isn't exactly how they'd organize these theaters then, but it's working for me). This is where he'd watch runthroughs of his plays, with his assistant next to him with a clipboard making notes. He loved himself best at this time in his life, felt closest to himself. He was back in the theater, where he'd seen himself when he was a kid. He loved the intimacy of theatrical control. He grew up in the small family catering business, with the show business urgency and flair where, like the theater, each scene is on and over in an instant.

My mother is at the piano, in the orchestra. This part is easy. The part I work on every day involves placing and shifting other characters as the day's play changes.

In a classic memory theater, the values you need to recall are placed in arched alcoves around the courtyard, virtues being more important players in your life than human characters. And they would last longer.

You store the thing to be memorized at the base, like in a little drawer the statue is standing on. Or you can put it on a platform, a slope or stairs that climb and pause. I could start with the loft above my mother's studio where I had a place to paint and write. Steep stairs, not so easy to slip up to and invade. The other stairs were the spiral staircase from the projection room to my father's bedroom—so he could work in bed when his back was bad and studio people could visit without disturbing my mother. There were back stairs so we could get from our children's wing to the kitchen and the breakfast room. I had a class problem about being a kid. I liked taking the broad front stairs and going through the dining room, then through the swinging door to the kitchen with some clear authority.

After Judith leaves, I go down to see Laurie. On the way, I glance in the mirror on the stairs.

Laurie opens her door. I hug her, admire her new picture, then say, "Judith says I've been looking in the mirror a lot. Is this true?"

"You probably don't have a strong sense of yourself right now, so you're looking to see who you are—here—to see what you look like." Laurie gets a camera from the bookshelf to the right of her easel. "Just look at the camera."

"Don't take a picture. I have nothing on. You know what I mean. Makeup."

"Be quiet."

The picture is taken.

Does Elizabeth Hurley have a place in my memory theater? After I have tea with her tomorrow, will I remember where I'm coming back to? A whole new fear. Maybe that's why my rooms are so busy, so bright—if they pale out in memory while I'm away, there'll still be something here.

Liz Hurley doesn't meet me at The Ivy, nor in the satiny boudoir dressing rooms old stars liked. She decides to meet in her office, a film company set in a couple of floors of an old Victorian house, large whitewashed rooms with giant posters of Hugh Grant propped up against the walls, the way you place paintings you may not keep.

She clomps across the bare floor in wooden high-heeled clogs with brass studs, suede vamps in Jil Sander palomino pink. She folds herself down into a deep yellow and white sofa.

"Until this year, I only had running shoes, everyday boots, black high heels. Now I have masses of high heels, beautiful high heels," she says. "I always wear monster heels. These are low for me, slippers."

"Do you know," I point out, "Rita Lydig, who wore a tricorn hat and was sculpted by Rodin and painted by Sargent and Boldini, had three hundred pairs of shoes?"

Something she's been just waiting to hear.

"That's almost as many shoes," I continue, "as the tartes tatins Bill Hayward made last year. Bill made one every day. (Have I told you this?) He's Brooke Hayward's brother—their sister and mother died when they were young—Brooke wrote the story in her fine book *Haywire*. Now I realize Hurley does not know Bill Hayward's dark legacy, and doesn't want to. This is like coming to London and not wanting to know all about the places with the blue plaques. You can't really be in the movies without knowing all our family scenarios. But what beginner wants that. What all she wants is that we should know her.

"I had an hour's sleep last night," she continues, "I look absolutely awful. Really dreadful." She's wearing dirty white jeans with button studs up the front. Her high breasts shimmer under the iridescent silver threads knitted into the bare midriff sweater.

Cecil B. De Mille told Paulette Goddard when she went out looking ratty, "You're a star. You never even cross the alley unless you're dressed to the teeth." Stars, my father said, always need to be told they look terrific, that they're doing a great job. So I won't agree that she looks dreadful.

"I would like to look this dreadful," I say, then, "How did you come across the Versace dress?"

"A friend told me someone had a rack with PR stuff, to 'go have a look,' so I went and I said, 'Um, yeah.' There is a problem of having no hips, completely flat from here to here. But last summer Versace cut a lot of little suits to give me a little wiggly bottom and hips, which I wore with Estee Lauder spray-on tanner—nicer than nude tights. This fall everything I've chosen from Versace's couture line is winter white. Dare I wear it?" Suddenly she's Pamela Harriman, statesmanlike responsibility to Fashion.

"So I asked Liz Hurley, 'What sign are you?'" I report the following Sunday after I finish reading the story I've written to the group.

"It would be difficult to be from L.A. and not ask this question," Angie agrees.

"So what was she?" Geraldine asks.

"Gemini. We were thrilled with each other for an instant."

"Then, only because it's next to 'what sign are you' in that kind of question, I asked, 'Where will you be on New Year's Eve, the night this century changes?' and she said she'd probably be sitting very sadly, 'by myself, on a foreign film set.'"

I am looking again at Laurie Lipton's picture of the slight woman with a steely glance and a vague resemblance to Alan Rickman. The woman has a pen in her left hand, a writing pad on her lap and is sitting in front of an intricate theatrical vision, an imaginary movie studio set, with swags and floodlights, stars and crew—and there, in the front, a left-handed little girl making notes.

Laurie has given me the perfect memory theater. Everything I love is here. And the woman here is okay.

But these aren't the sharp, tough, critical people I've heard I once knew. So who did I know? I remember my brother and sister and I dressed up to meet Easterners, senators, moguls, stockholders, theater owners. I always told them I was going to be a writer. So I was a kid, three feet high—you never know. I damn well might do that. We wouldn't want you growing up remembering us in a way we wouldn't like.

Maybe this was a way I got you to pay attention to me when I was a kid.

The next morning I mail off the piece on Hurley. When I come home, I have a message from Professor Rose, who has agreed to see me when he returns from a lecture tour in America.

It's as if the characters are writing me back. It's not much, but it's a beginning. This is the book I am writing. The dimension I am in is the present.

23

Steven Rose began studying at Cambridge in the late fifties, when he discovered what he calls "the exciting hinterland between chemistry and physiology" and named the science of neurochemistry, the biochemistry of the brain. Rose is now a professor at the Open University an hour north of London.

In the car, whizzing past the industrial plains bordering any freeway, my body's clock forgets where it is. It careens in and out of time zones: New York, Los Angeles, or here. Where is here? I take a quick look at my datebook.

My day is a charade—signs are held up, expressions, gestures tossed at me: Who is this? Where is it from? What does this have to do with that? Get a grip.

The university spreads flat and starkly threatening, like giant blocks, across acres of flat land. I won't remember one corner, one path beyond the huge white reception building.

Don't have to. Someone is meeting me. A positive approach is not to worry about remembering what you don't have to.

The professor's assistant leads me down a bright hall to his office. Steven's standing, instantly engaging, with wide, enthusiastic light eyes.

He's good at capturing eager attention, but at the same time, you sense that he'd rather be immersed in his work.

"So, baby chicks everywhere," I say, "I'd know it was you."

There's a toy windup baby chick and a kind of wooden paddle toy with several chicks. "The Ukrainians bought that for me," he says.

One of the things I liked best about his book is the sense you get of the man writing the book, of his own experience with memory: being in an air raid shelter; the childhood artifacts, like a silver mug, that trigger his own memories. I stretched harder to understand what he was talking about.

I'm looking him over so I can remember how to describe him. People are not used to someone casing them so carefully.

"I'm sorry I'm dressed like this," he says. He's wearing dark slacks and a dark tie, and the soft deep ultramarine shirt fits his torso well. You expect him to be heavier because of the warmth of his writing. But given how much he does, he doesn't have time to eat. You also expect him to be in baggy khakis and the mindless patterned sport shirt that thinking men live in.

"I have to go to one of those dinners tonight," he explains, "so that's why I'm dressed like this."

"It's okay," I say, "I won't think you're intimidating."

There is a desk, a computer and a round table with royal blue chairs where we sit and talk. In the center is a large square fish bowl filled with light gold water; inside is a human brain, which looks smaller, trimmer than I'd expected. In his book, Steven talks about doing his early research on kosher cows' brains so that when the work was finished, he could use them for his dinner. I will not ask if this one is marinating.

"I have some questions, but the embarrassing thing is they haven't stayed arranged in my memory theater. You've talked about Hyden and how cells change their properties as a result of experience; that large cells deep in the brain concerned with balance don't change . . ."

"That was back in the sixties." How fast he reaches back and connects. But, to me, with time unfixed slipping like Mercury, the sixties are yesterday, and I only know the eighties by that red Valentino suit.

"Now, this is the question I've underlined here—I've got four stars next to it."

"I see that," he says.

"I was wondering if under a large shock, say a grand mal fit, would the structures react and reset so radically that the whole character of the memory changes. One of the things that has surprised me the most, particularly when I go back and reread things I wrote before, my attitude and angle have changed, and my appreciation of the memory and understanding—now, this may be age or maturity—but it seems that some things are just radically different."

"It's a difficult question, isn't it? I have the experience of reading things I wrote fifteen or twenty years ago—firstly, I have no recollection of writing them at all. Secondly, I look at this thing and don't recognize it as mine, so I don't think that's as it were unique to a post-fit situation. In fact, my partner, a sociologist, says exactly the same thing." He speaks of her often and with affection, and I'm thinking how much more I'd like to have them over to dinner than to learn how my brain works. Or, to be precise, if it still does.

He smiles. His eyes light up when he's onto his subject; it's a physical sign of enthusiasm. Does it show in the neurons and so on, or only in the eyes, to be caught by untrained eyes like mine? Ed Doctorow's eyes lit up like this when he'd talk about a character he liked. You'd see this electricity in the eyes of some movie stars in the first dailies on a new picture. And always with Fred Astaire, always Gene Kelly.

It's his fresh involvement that makes me reach to catch what he means. One of the things that happens after an epileptic seizure, electric shock treatment, and so on is you get random bits of nerve cell death. "On the whole," Steven says, "they don't get replaced in adults, so, in a sense, there're going to be spots that are disconnected. We know a great deal more about the mechanisms involved in learning than we do when we try to remember something, as it were, inside your brain. The mechanisms for scanning and refinding that memory are almost entirely unknown and uninvestigated. We still don't know what changes in the brain to create memory storage. We know a change is required in the biochemistry and structure of cells, but we don't know where this happens, either."

"But you said there are medications now that can improve memory." I don't forget this idea.

"Yes, but the complete change and adjustment of memory is not likely to happen with a pill." He's careful. "That would be like trying to fine-tune a radio by jamming a screwdriver into it."

I see it as fixing antique lace: as tying and threading small connections, picking up beads, examining them in the light to get each shape and shade in place. Once it's here, it looks together and may even (if you don't go too close) look like it used to, but you mustn't confuse it with the original. Make the same demands and it will shred apart into pieces, as airy and dead as cinders.

"There's a whole literature about so-called 'smart' drugs," he continues. "Most of them don't work. Some of them are designed for people with Alzheimer's disease and I think they're pretty tough to take, and I wouldn't advise taking those at the moment. There is one substance that is used in Europe—it's not licensed here—that is completely innocuous, it does no harm at all. It might well be worth trying in your situation. It's called Piracitan—I know the company that manufactures it, and we've used it in the lab. I don't think it would do any harm to try it. You can order it off the Web."

Before we go on talking, we go to the commissary for lunch. I am reminded of school cafeterias—and the commissary at the studio where an actress would have a dish of cottage cheese, and you'd see her watching the director eat a big plateful of goulash and noodles. I have chicken and tomato sauce with pasta, and he has a herring and onion salad with a baked potato, which looks like a far better idea.

I love institutional lunches. I have the big white visitor's badge clipped to my sweater and I feel I am involved. I do belong. Instantly I look around to see which rank I'm in. Wherever I go, I'm figuring out a way I can muscle in, get a gig. Is this because I don't relate in any gripping way, once I am not there, to any place?

We're going back to his office now through a labyrinth of walks and hallways.

"I'm glad I'm with you or I'd never get back," I say.

"Not being able to find your way around is one of the things I would expect to happen," he says, "a fairly logical step if you've had some sort of disconnection.

"The part of the brain called the hippocampus is associated both with short-term memory and the transfer of short-term to long-term memory, and with spatial representation, that is, making maps of where you are."

It's as if here I'm reassured. I know exactly what he means. "I not only don't know, but I don't know I've been there."

"But I bet you're unimpaired at riding a bike," he says. "That's procedural memory. People with Alzheimer's disease, a progressive loss of memory, can remember how to ride a bike, long after they may have forgotten a bike is called a bike. The first thing to go is autobiographical episodic memory—'Did you see that film last night?' To know that a film is a film—to remember the days of the week (semantic memory) is easier—but to know what did I do last Tuesday is harder."

I think about that. I don't.

"What interests me is that my writing is off. I have trouble remembering sentences or who a character is, but my drawing is better than ever."

I'm not clear again on what Steven Rose means when he says, "It is always declarative rather than procedural memory that suffers."

But I woke up late one night, went into my workroom, and decided not to try to write, but to send something to Phoebe. I picked a tiny notebook and I sat there with the colored pencils, filling pages until the book for Phoebe was finished. I remembered how to shadow, to feel where the line's going before the pencil touches the paper. I remembered the strokes and curves to take to bring coyotes, bears, and starker characters to some sort of life. I remember learning from my mother. But I don't remember where she died or when.

"In order to write now," I tell him, "I start by sometimes drawing my way in. It's almost like I'm fooling the brain, relaxing my brain so it won't think about it."

"If I were a clinician listening to that, I would begin to suspect there was damage to the left hemisphere rather than the right hemisphere."

"That reminds me," I say quickly, "I wanted to talk about the left-handed rats and the right-handed rats."

"Those were very early experiments," he says. "The man who invented them influenced me and shaped the direction of my research back in the sixties. He trained rats to reach for food down a glass tube and noted whether the rats reached with their right or left paw. Then he made it necessary for them to reach with the opposite paw. It enabled him to study these particular cells and that particular region—so he designed his task to be related to the certain cells he was studying at the time. Of course, methods are much more sensitive now and those techniques are completely unused. Humans are the only organisms that are not lateralized; there's only a small proportion who are left-handed. If you look at other animals, about half prefer their right paw and half prefer their left paw. However, that's not true for birds. They have a very sharp lateralized brain."

"And a good sense of color." I read this in a book by Barry Gordon, an American. Gordon is a psychologist who works directly with people.

"Is memory DNA?" is my next question. A man named McConnell wrote about this issue in the *Worm Runners Digest*, which was not a journal for fit worms, but a study of worms who had been trained by light, then cannibalized by other worms to see if they'd remember the game the food worms had been taught. It seemed that the worms who ate the light-trained worms did change their behavior as if they remembered "the conditioned response their food had learned." This created quite a stir for a few years, until, as worms do, they turned and refused to duplicate the experiments for other scientists.

Steven gave me a book for children called *Brainbox,* which he wrote with a twelve-year-old friend of his. He'd asked his young friend what he wanted to know about the brain, and being a very democratic young man, the twelve-year-old polled his class and came back with a list of fifteen questions, of which the last two were, "How does your brain make you happy or sad?" and "Are you the same as your brain or different?" Steven was deeply impressed.

"I think I'll think about that last question for a long time."

"I think many people have," Steven says.

"I'm wondering, is recollection shadowed by emotion?"

"Oh, yes. Why do we remember things? Not in order to have cognitive skills, to be able to remember the letters of the alphabet—but to be able to survive in the world. To survive in the world is entirely about emotional rather than cognitive things."

"What I fear I've done," I say, "is reinvent rather than recollect."

"One always does," Steven says. "Every act of memory is a reinvention, isn't it? When you tell a story or have an experience, the memory shifts as the tale is a little different next time it's told. Its character also changes depending upon who you tell it to."

"And how you're feeling as you tell it this time."

"I know that when I tell a story out loud, it changes as I adjust my own memories to my listener's response. I never tell my father's stories or my own exactly the same way to everyone. Times and attitudes change, subtle language changes restyle a memory.

"That's why this whole debate is going on—the psychotherapists' memory-false memory debate in the States. People are supposed to be able to remember sexual abuse or whatever when they were kids. So if I asked you to remember something now—which is your favorite child-hood memory—you'd be remembering it as you changed it the last time you remembered it. Once a psychotherapist has persuaded you that you were abused as a child, that memory's real to you—it's in the brain. It's as real to you as if in fact it had actually happened. We all tell ourselves our stories. Memory isn't intended, as it were, as an absolute record. It's an active process going on inside the brain."

I am lying in our bed in the soft gray morning light. I am not going back into old time. Stay in a place you know here. Here is London. I am in Regent's Park. I'm walking along the path to Venice. Is that what it's called? And I'm in a gondola. (This may be on a trip to the real Venice, not London's Little Venice. Not L.A.'s Venice.) Did we go to Venice with Judith and Cesario? Yes, and they fought. She flew home, Prada bags on her back, slung over forearms, and swinging from her shoulders.

Never mind. Now I'm carefully guiding this gondola around rivers of nerves through my body, bumping along by the joints, hearing the vibration of the veins like the shimmering of leaves. This once led me to see that I could stay in touch without drugs; that, in fact, they muzzled the connections. Since then, I can connect but I can't see the memory. I've never been able to find it.

I will respond and connect my spirit and my mind to my body. And it will work. I think of Professor Rose's floating brain. Do I remember the part of my brain that writes? Can I image the connections to fix them?

Possibly. Steven did say that when neurons, cells, die, it gives more space for connections with other cells to grow—so it's not necessarily a bad thing. In the first ten to twelve years of life—our development is really rooted in these first years—there's a huge overproduction of neurons. The brain is a very redundant structure—there are many pathways to get from one place to another—there's more than one way to skin a cat, so to speak. "So if some neurons died in your seizure," he shrugs, "you'll just find another way to do the same thing." I thought about Kohler's chimpanzees, who got the idea to put short sticks together to reach their bananas. When animals learn a goal in a maze, even if you turn it around, change it, they'll find the way to their goal.

Maybe scientists who do not work with people may take a more positive view of things. Working with five hundred rats and baby chicks that don't make it, it's the one that does that counts.

I'm going to be the one. I'm going to find another way to do hours and days—and scenes and chapters.

Steven Rose has a memory test in his little book, *Brainbox*, which I've adapted, changing some of the color images so the drawings will be clearer in black and white. Look at these things for one minute. Then see how many you remember.

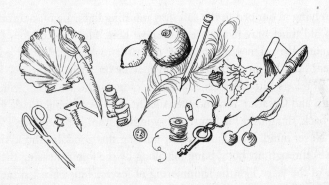

This is what I remembered the first time I took the test: lemon, orange, feather, safety pin, pencil.

"*T*heater is memory," John Lahr says, "that's its function. Theater is an historical memory; a way of taking an audience and making it imagine the past, a way of working through the past, a way of reviving memories to reflect on their own lives. All of Shakespeare's history plays help an audience recall in themselves parallel universes, see if they can find those feelings in their own stories."

John eats with the busy vigor of a historical character, smiling, eyebrows as alert to flavors as to his own brisk delivery. John loves to eat. Loves to think. Loves to talk about what he's thinking.

I'd called John a few weeks after I saw Steven Rose. I had John's number in my book, and I knew we'd met. Or maybe I'd been given his number because John and I have something of a similar background. I think. He's the theater critic for *The New Yorker* and his father was Bert Lahr, the Cowardly Lion in *The Wizard of Oz*.

"John . . . Hi, we met years ago in L.A. I can't remember if we went to school together, but . . ."

"Jill . . . ," he was amused, patient. I loved his voice right away. Now, is this because I remember its character, or because from the sound of his voice I know it's one of those which easily tumbles over into laughter. "You were here in Primrose Hill a few months ago."

"Not the Upper West Side?"

"No. You got lost, but not that lost," John says. "Why don't I come over—let's talk about memory."

"I'll feed you," I say.

"That's what I hoped," he says.

"The first time we met," John explains when he arrives, "was when our books came out—*Notes on a Cowardly Lion* and *Bed/Time/Story*—and we were on this show as the children of famous people. But we really met when you and Stuart came over to England, your friends and mine, Leslie and Arthur Kopit, gave you our address in Hampstead. You'd arrive for tea with muffins, always tied in red bandannas, and jalapeño jelly."

"No—not muffins—chili corn bread—yes. Or Tex Mex scones!"

"So you remember food?"

"Far better than anything."

"You struck up a relationship with my son, Chris. You were very nice to him," John says. "You always brought either strawberries or a strawberry tart just for him. You'd just come and schmooze with everybody, and then you'd call for your car, which I thought was rather magnificent—that you had a car. Then the guy who made *Four Weddings and a Funeral* was giving a party and you and Stuart were there.

"Do you remember where our teas were in the house?" John asks. "They were at the top of the house. Once you got to our house, you had to go up another set of stairs."

I lay out lunch and John sits down. "Of course," he says, "theater is memory."

"When you lose your memory, you don't think you're very smart anymore. Being without memory in a territory of writers and artists is like being in L.A. and not driving."

So my idea that memory is theater and John saying theater is memory restores my self-confidence, which is why I'm sure, when I go over what we talk about, I see that I've remembered more than usual.

Like me, John started out knowing he was a writer, because "I could do that better than anything else. I got a lot of pleasure out of it." Like me, he began with political journalism.

John was at Yale when they integrated Ole Miss and they sent him down there to cover the story. "They were rioting on the campus, it was all aflame, and they had banned all reporters because someone from *Paris*

Match had been murdered. But I put my secret society pin on (it just happened to have a chapter at Ole Miss) and I was able to walk on campus as a student. I scooped the nation for one day and won that year's Yale Writing Prize."

How quickly memory shifts from our own lives to our parents'. "My early desire to be famous came," John says, "from my parents. That was all they respected, all they knew, and it was the world I lived in. I didn't know anyone who wasn't famous."

We've dined out and well on the material they are. But memory is what all writers use. It's the fabric; the trick is how you cut it.

John slices another piece of challah, and I wonder if I know a writer who does not eat much.

"It's a very strange and unsettling feeling thinking of your parents not knowing—intellectually," he continues. "I was always interested in words and in being able to use words, and I think that has something to do with what my parents didn't have."

"I think it can go two ways. My father loved words. He made them so real you could almost see them, like characters and, like you, John, at the teas, he was a generous presence. He made sure you got to do your own riffs, unless you went on too long. He liked quieter writers and actors like Robert Ryan or Richard Widmark and doctors and political friends who could write their parents in Detroit, Cleveland and Philadelphia and say they'd arrived. Doctors picked up the style of the storytellers and each week was like an *ER* episode. They'd battle not only over who had the most famous patients, but which one was in bigger trouble.

But my father's favorite friends were academics. They were famous in their own universities so, I think, he felt they gave him a certain exclusive credibility. "When he had Abraham Kaplan to dinner, this almost made up for not going to school. This is not unlike me liking to write about going up to Oxford, as if I can catch some of that, it's as if I can say my brain has arrived. The way people go to Hollywood to catch stardust."

John says, "I wrote a novel called *The Autograph Hound,* about a guy who pursues famous people, and I was trying to explore those feelings.

"I'm embarrassed I don't remember it."

"I wouldn't expect you to," he laughs. "Even after I've written a book, I'm getting on with the next thing, so I've completely forgotten that information because more information has come in. I don't know anybody who rereads their books. The book represents a little time capsule of who you were in that period, what your concerns were, even stylistically.

"So we do change. Maybe it's true that you can't really write for too long from who you really are. You start out as a basic, clear and wonderful creature. Then the memories pile up and the pressures of who you ought to be now, the pressures of the time, your physical condition and the pressure of how much money you have—all these start pushing and shoving this talent's morality; and as you work through life's struggles, it affects how you express yourself. You change your 'look,' as it were, because you are thinking about what you want to project. You can't do that in writing—you do see people do it, but they're very bad writers who want to project an image. They want to be Henry James or Roger Burbank. What it communicates is their emptiness or their confusion. It doesn't let the reader in—you're trying to let the reader see you in as honest a way as possible."

"Do you take notes at the theater, or do you remember a play and moments or lines?"

"I do take a few notes, but I like to applaud. Most critics won't applaud at the end of a play—they're not part of the experience. I go out and rediscover the play over the next couple of days when I'm analyzing it. Very recently, I went to see *Electra*. I went with a friend and afterwards we were talking about the play and she said, didn't you love the moment when the servant picked her up in his arms, and for that one brief moment she fell asleep suddenly, the only moment she was able to relax? And I said, what are you talking about? I didn't see that, what do you mean she relaxes? What happened, I think I was writing in my note pad—I missed that moment, it just didn't exist for me. So I went back to see it again—it was a very good play. Having seen the play once and having gotten to know the parameters of the play, I felt very comfortable with it. I could listen to it in a different way. So the second time through, I not only saw the moment my friend had mentioned, which was a very important moment, my opinion of the play was clearer for me—I was completely in the play.

"That's one of the ways I know if it's a really good play or not. I've had some wonderful times at the theater when I was never 'at the play.' If it's not fully dramatized or if it's boring, I'll sort of go off and follow something in my own head. I'll make something up or I'll be thinking about something before I came to the theater and I'll just time travel and dream away. That's another reason I don't like to take notes. The fun of going to the theater is you enter that world and stay in that world."

"I'm not sure I enter my life," I say. "I live as though I'm saying, how am I going to write about this? how can I use this character, where? So much of my life is . . ."

"Scavenging," he says.

He's right.

"Yes!"

"You are, to a certain extent, using your memory predicament to scavenge yourself. There are a lot of people you can go back and say, what happened? What do you remember? Of course, the sad thing is that what one person selects may not be the thing that you may have selected from a particular occasion."

We start to talk about John's teas, which are part of what little I remember from my first years in London. "They reminded me very much of Friday afternoons in New York, when the *Vogue* editor, Leo Lerman, had us gather in his office. He reminded me of Monty Woolley in sharp thirties roles. I think I mean the thirties—but, yes, that's where I met Joan Buck. She was also a young writer and I could feel her excitement. You'd come with your lines sharpened like pencils. Leo was like a professor. If you were fast, he'd cue you in like this week's star."

There was a little of that feeling at John's teas. Breathless before you'd talk—is this a smart angle? I don't say that now, because in memory's favored soft light I felt at home. The teas were more like our Sunday dinners.

"Remember, there was only one rule," John says, "which was that nobody was to write about those times."

"Because I will write about it . . . or, as you say, scavenge." Which I have already done.

"No, it's okay," he says. "For me, too, it was a non-self-conscious time—it was a way to be with people—'cause we were solitary a lot of the time. We did the teas for a decade—we never knew who was coming, we never invited anybody, they knew it was there, they knew it was happening. The teas were unvarying. Never, in ten years, did we have a Friday when no one came. We provided the tea and everybody brought something. One summer, when we left to go on holiday back to America, they didn't want the teas to stop, so we left the teas with Francesca and they had tea at our house even when we were away! We had the brown tiles on the floor, sort of Provençal tiles, and a pine table. It had a very American feeling to it."

I remember it had, to me, a very English feeling, like what Josie and all of us wanted to feel we had in the Village, as if we'd just come over from Oxford and started our first novel. But I am probably remembering what I want to.

"The teas would start at four-thirty and they'd end when they ended—Steven Spielberg could be there, Richard Avedon, anyone who was passing through. It was like a novel—everything happened over that table over a period of ten years. Lives changed, people separated—in fact, in the end, Anthea and I separated. I carried the teas on for a while until I got *The New Yorker* job and it got to be too much—plus, it had been part of our old life, not my new life. It was a lovely memory. When Anthea left, I put a lot of money into the flat—redecorated it, completely redid the bedroom—went upscale. We had been married twenty-three years. I wanted to lose the memory—instead of regain a memory. It's odd in that sense. In those twenty-three years, I was only away from the house for five nights. If you ask me now what I remember—I do remember things to do with Chris—I don't remember a lot of it. It's surprising how you can will that stuff away. There are some scenes I'll never forget, especially at the very end of our marriage—like Bergman scenes. Sometimes when I'm waiting for Connie at the end of the day to pick me up, I'll stand at the window—and just when I'm doing that, I remember Anthea waiting by the window, waiting for her soon-to-be-husband to pick her up, and I find that very painful."

"You talk of standing by a window. My parents put a window seat in this window where I'd look out over the driveway, and I'd sit there, watching my father leave, waiting for him to come home. He was the center in the house. The captain. When he was gone, I felt we'd go aground.

"I think I was sitting in that window watching my father drive off at three-thirty one morning. I'd been up reading. The medicine they gave you for asthma then, ephedrine, was like speed, so I heard the phone first when it rang in the hall off the children's wing. It was Monty Clift calling and he sounded terrible. My father sped off too fast, down the drive, to see him, but by the time he got there, Montgomery was dead."

"Didn't your dad make a picture with Clift—Nathaniel West's *Miss Lonelyhearts*—much later?"

I think that over. "Of course. Then I've seen that wrong, that was *Raintree County*. And Monty did survive." I remember the call from a different time.

What I remember is the call, the voice, the fragility, and fear. And my father's concern as he rushed off. But Monty didn't die that night. And all these years I've remembered it that way. What I probably saw from the window is my father rushing off the night my grandmother died.

How important it is to logic memories through whenever you can. The sadness I felt that night, the worry about my father, was mixed up with my grandmother's death.

"You know," John says, "Bergman makes no distinction between the past and the present. The past isn't a shadowy ghost, and the people in the present time are in constant conversation with people in the past. I think what writers do when they close the door and start to imagine a world is to be able to call out of themselves these ghosts and they become very real and alive. I've had one dream where my father appeared to me—I could smell his cologne, I could feel his cashmere coat—it felt incredibly, deeply real. No writer would not admit to talking to themselves when they write—with these figures of their past who haunt them—in my case, it's my mother and father, and Anthea—although both my parents are dead. Writing is one way, one socially acceptable way, of communicating with the dead. My mother wanted to kill certain memories, so she just wouldn't tell us. We are left with very little about her early life. Memory gives a sense of coherence to a largely incomprehensible world, just to make a pattern so you can feel at home in the world—and if you don't have that, it's a frustration."

I tell John the story about Stuart's mother. No matter how ill she was, as we'd turn the corner up to her house set so alone with the bleak indigo moors stretching out on all sides, she'd be there, dress shuddering in the wind and her sweater shawl round her shoulders, waiting outside for him. My memory of the story changes now as I tell it to John. And my memory of the moors—sometimes they're lavender and embracing. I tart up the story as some writers do when they're together.

John tells me now that a few weeks ago, he came across a conversation he'd had with his mother when she was losing her memory. "Although there was a certain essential 'momness' to her, actually, without memory, you lose a sense of who you are, a sense of self, because really the self is only a collection of associations and memories of things done and done to you and where you've been—and without that, it's very hard to get any definition on who you are. In the end, she could only speak about one or two sentences that she could remember. I used to go visit her in the end—and it's hard to visit, knowing that when you close the door to leave, she won't remember that you've come. She was in a constant state of feeling that no one visited her—of waiting for visits she couldn't remember. Memory allows you to have relationships with the outside world. You have to remember who's talking to you and something about them in order to engage them. Mom's last spoken words to me were—'You think I'm stupid. This house is closing, now and forever.' It was extraordinary language—not at all like the rhythm of the few sentences she regularly spoke at the end of her life—she had left. But the experience of being with her when she lost her memory—there was nothing to share—she'd lost her identity."

We're picking at tarts, a bit of cherry, a touch of apple. We keep eating to keep talking—I'll have another bite if you'll hear me.

Before he goes, John says, "When you write about memory you'll be saying, 'I lost my memory, and you're seeing me trying to retrieve, and here are some of the things I've found out.' That's very powerful—it's like a mystery—it's really clues to yourself. It will be powerful for a reader because you're so vulnerable, it's upsetting, embarrassing, shameful, distressing, the trap that memory has for people. That's why we forget certain memories, because they're too horrible to remember. We doctor them."

"I could say it's all fiction. Lay it on a character."

"That's a disclaimer. You can probably get Frances Yates's book on the Internet," John says, "it's very simple." He sits at the computer . . . and there it is.

Then he takes another cookie. "These are excellent," he says, and he is gone before I can say I didn't bake them.

The day after John has been here, I can't remember what we had for lunch. "You had a good rice dish," Stuart tells me, "roast chicken, and a raisin challah from Villandry, which you ate three quarters of."

"That's all right. Like the chili corn bread, I'll remember it in another eight years or so when it's turned into long-term memory."

25

I keep finding myself doing things that I remember I don't do. "Here I am at a symphony," I say to Shaw during the intermission at this stark concert hall on the wrong side of the Thames. It feels like the Valley.

"We've been going to London Philharmonic concerts," he reminds me, "for around fifteen years. We have the same seats every season, center aisle, fifth row."

"That may be why all of the players in the orchestra are as familiar as the faces on my kitchen wall," I say.

The fourth bass player who looks like Errol Flynn. The Brazilian cellist, the only one who talks to the young fat cellist. The tall blond is lead violinist, wild for Kent Nagano, the conductor with the long black hair.

Stuart comes to hear the music. I come to continue the soap I've made up about all the musicians.

"That," he says, "was my plan. Patterns build security. You liked your childhood because there was a routine."

"I hate habits."

"No, you don't. You like the meetings you know. And when the writing is hard, you wear your Sundance sweatshirt every day."

"I do?"

He's right, which I do not tell him.

I don't like classical music. My mother once said, and I remember this, "It's vengeance against me for making you take piano lessons."

Did I also tell you (and how many times) that I hate walking? And now I am walking with Luise Rainer after a visit to our friend Corinne's today.

"I saw this viola player from Russia last night," I tell Luise. Perhaps it was the night before. But I remember his expression. This is progress. "I can't remember if it was the dark pull of the viola or the fling of his long black hair that has caught me so."

"It must be the music," she says. "My first memory is of listening to my mother play Beethoven. I remember every note. I was two and I sat under the piano. Much later, of course, Gershwin played at our house when he stayed there, and Marian Anderson, too. I loved Rubenstein, but he blew so with his nostrils when he played."

Luise's voice is deep, Austrian. Under her fur coat she wears gold lame knit sweaters, tights and caps. Does she dress like a miniature Oscar to remind us she won two in a row, back-to-back for *The Good Earth* and *The Great Ziegfeld*.

I also sat under the piano listening to my mother play. I remember her red toenails, not the music. The polish was Arden's Victory Red. She pressed her foot down and down, again, harder on the pedals in her blue maribou mules. I wanted to slide my fingers in between her soft toes.

Luise is so tiny, so slender she's translucent, as I remember my mother was at the end of her life. I must not confuse them. Luise is electric and demanding. Asking for what you want may be longevity's real secret. And running. Luise was the fastest runner at her school in Hamburg.

"Centuries ago," Dr. Rose wrote, when he explained the memory theaters, "special people, the elderly, the bards, became the keepers of the common culture, capable of retelling the epic tales which enshrined each society's origins." The survivors, like Kitty Hart and the photographer Eve Arnold, whom I watch as a trip guide into the 'third act,' seem to operate on curiosity, on a dogged, exasperated intrigue about life right now. Nostalgia does not play a huge role here. It keeps its place. But Luise is different.

My mother told me after I asked Ethel Barrymore if she'd met Queen Victoria, "You don't ask someone really famous to talk about the past because it's like saying you aren't famous right now, and the only thing that makes life bearable is being famous."

What matters now to Luise seems to be walking and reading eccentric new books. But I still want Luise to tell me the stories. I'm counting on her to bring back that time, but she's living now.

She lifts her sharp jaw in what I see as a period expression. My mother did this. It comes out of the scenes when the star (Luise) would look up into the man's eyes. "Memory and the soul are one," she says. "I am the soul. The body is borrowed. I have been lucky this time—it is a wonderful one."

If the soul is a matter of faith and if the memory and soul are connected, one might have more faith in memory turning up when you really need it.

As we cross the park, Luise turns up the collar of her coat and becomes, with strides, the officer she's telling me she was in the Middle East during the Second World War. "I was married then, too. I miss him some days—more often." She stares across the park as if it's a stretch of time. "Of course, I do not want to be alone. You say you will come and have lunch again. You say you love me and yet I never see you."

How can I say I forget I know you? You cannot forget a legend.

"So, I saw your picture with your friend in *The Evening Standard*," Fran Curtis, a friend of Stuart's who had us over for dinner, says. "Jill *knows* Streisand." She says to this other couple she and Roger have over.

I can pretend I have forgotten what Fran's talking about. When you are friends with a star, what you have to offer is discretion and privacy.

"Oh, yes," I say, "this is great smoked salmon. Is it Selfridges's?"

"No," Fran says, "I went to Villandry. You told me they have the best. Don't you remember?"

We always say smoked salmon now. It was lox in L.A. and only my mother and Mildred Jaffe called it smoked salmon. It's the easiest and most expensive starter here. I would rather eat the plate.

Or anything to avoid the discussion Fran is determined to have. I can see now why I have been invited.

The other couple, Abigail and Ben, are London establishment. I knew when I opened the dashed-off little white card inviting us to dinner, handwriting thin as Fran's stiletto heels, that we were being asked for a purpose. "Do you see Barbra every time she comes to London?"

"Sometimes. Is that a Jil Sander top? I like it."

"Yes," Fran lifts her shoulders, "it's sweet, isn't it. I bought it in L.A." There has to be a reason to say one bought something in L.A. "Our house in Malibu isn't far from your *friend's* house, you know," she says.

I know exactly. She'd like to meet Barbra.

"So where do you live?" I cut right across and ask Abigail. In London, that's a little fresh, asking 'where do you live' before you know someone really well. But since the husband, Ben, owns real estate, I feel it's okay.

"We live upstairs," Ben says.

"That's how we met," Fran says.

"And after the smoked salmon, I'm going upstairs to work," Ben says.

"He means," Abigail says, "that he's going upstairs to drink."

Fran winks at Stuart quickly. Maybe she thinks he'll talk to Ben about drinking. Second on Fran's agenda.

"He walks everywhere," Abigail says, "you could go together."

"There's an idea," Roger says. He and Stuart nod at each other. We all know this will not happen.

Fran says, "I wonder what a movie star does when she wants to go window shopping. Personally, I like just walking with friends so much."

"Me, too," I say.

"That's absolutely *not* true," Stuart says. He picks up the last chocolate from the plate.

"I'd love to come to your writers' group," Fran says before we leave. "I'm writing a wonderful screenplay. So what do you all wear?"

A movie star does not go window shopping. This star has enough trouble shopping.

Fran was referring to a photograph in the *Evening Standard* of me behind Barbra, leaving a London Store. This happened a while ago. Barbra had seen something at the restaurant Le Manoir aux Quat' Saisons, which the maître 'd, Alain, had told us we could find at John Lewis.

"But it's like Macy's," I say, "you'll be mobbed."

"This will be different," Barbra says. "It's England, and we'll slip in and out." Her tone of voice makes it clear we're going. She sends her limo to pick me up. Fame moves through a crowd like one of those dog whistles you can't hear, like the sonic control or whatever it is that sends messages without wires. Stars have it. They are satellite personalities to a major degree, like the one in your family everyone gravitates around, only more so.

But Barbra is right—it *seems* different in England. As we shop, everyone notices. The difference here is no one says a single word, no one comes up to her with a card to sign, a song or movie idea. Just enough excitement flurries the staff into action.

One very cool aspect of shopping with a star—everything is possible. Immediately. If I wanted to do star for a day, it would probably be to get things fixed and delivered fast.

Although the English shopping public may be more reserved than we are, the press is just the reverse. A manager suggests that the car be brought around to a service exit. But by the time we get there, the press has found the same door. We burrow through the reporters and the snapping lights until we slip into the dark silence of the limo's back seat.

"I've been famous since I was a kid," she says, with exhaustion. She says it like fame is a kind of affliction.

And suddenly I remember how it had always seemed that way to me. It always seemed like an emergency—watching stars being rushed around, scared into time panics by studio and guards, crumpled breathless into limousines, watched over and pampered like little kids, never just free to come and go with no one watching, no one knowing where you were every minute, no one inviting you 'just because.' But then, if they don't invite you, you wonder if something's gone wrong. I'd wonder about my father's last picture, whether the box office was okay. But then every kid worries if the parent's scene changes: will the mood change?

My mood changes when Barbra calls. Another time: I am in my changing room taking things off. I'm down to sheer black pantyhose and black satin shirt, which is good for going into Stuart's changing room. Then the phone rings and it's Barbra's assistant, "She's got extra tickets for the theater tonight. Can you be ready in an hour?"

I put down the phone, dash into Stuart's changing room, already dressing; will he change plans? Agree to go out?

I don't want to tell him how being with her somehow pulls up a long-ing, for home, makes it feels so strong. But he knows. And, he'll go. Says he won't, but he's getting dressed. Now, that is.

"The friendship is about memory," Stuart says. "Barbra likes to hear your memories about Hollywood. And sometimes when you're with her, you remember how it was to be with your father. It's like coming home to a time you don't quite remember.

"She is also interested in the way you and I get along. We are real for her because she wants to see the truth—what really matters. A lot of her best friends have strong marriages."

"Not easy to find," I say.

He shows me pictures of times I've forgotten, of couples gathered around Barbra's sedar table. Are we like bookends, holding up our stories to keep them straight, easier to dust?

Line up her pictures in order and I'd guess the roles she's picked have tracked the evolution of my options as a woman.

"Barbra expects you to write again, trusts that you will do it, always asks how your work is going—and listens to the answer. Her curiosity's intense, like a journalist or a researcher. 'What do you mean by that?' 'What do you do with this?' You like that she's curious about you."

And then, I like to see how stardom has evolved.

Stars like Grace Kelly and Ingrid Bergman, Judy Garland and Elizabeth Taylor were frustrated by not having the power to do what they wanted. They would have had to be crazy to say 'No,' or 'This is what I want. Do it again. Do it my way.' Write their own books, like Shirley, invent their own personas, produce and direct their own movies like Barbra? Choose their co-stars? It would never have occurred to them.

Barbra's success allows her to do that. She's never pretentious. Perhaps because there's nothing Barbra has to pretend to that she isn't already.

I remember watching Barbra in her cream suit, heels, and fedora, looking at herself in the mirror at Le Manoir aux Quat' Saisons, an English country house hotel. All independent authority, she considers, adjusts. She's the director and the star.

Quite another matter when I watched Grace Kelly looking in a mirror in Helen Rose's costume design studio at MGM in 1955. This star stood with the preferred deference of the time, almost a condition of the contract, manners. I sat sketching quietly, waiting for the fitting of my first wedding gown. Whether Grace Kelly's gown was for the real wedding to Prince Rainier or for the wedding in *High Society*, any appearance was a part, a job. Every move had to be fresh but not unsettling, endearing but never sexual, stylish but not original. Expressions must be sincere, authentic, but not extraordinary. A flick of the eye this way, a slide of the smile there, and it's off—you've disconcerted the camera. The exchange is cut, the photographer loses faith. The expression, the voice—once they're gone, you can't find where you are. Even in a real day you were on. You didn't have a real day.

So here I am, one more time, walking.

Stuart's taking me out because I'm depressed about the one thing he can't fix: family. He never even tries. Distraction, as he knows, works best.

Before we go out, Johanna calls to tell me that Justin really doesn't want to fly alone to visit us this summer. "You don't know how it is to fly alone when you're a kid. You never went anywhere alone," Johanna says.

"Neither did you. I was with you. I'm sure I was. Or Nannie was."

Nannie was the Scotswoman who stayed even when there wasn't any money, fended off the landlord, and drank with me all night listening to my latest pages. Great for me.

"No," she said, "that's not true. We flew across the country alone all the time when we were eight and ten."

"I can't believe that!" I hate this.

"Mom, you don't remember."

"It's true." That's the darkest area, so dense with shame.

So before the phone can ring again, we leave the house. We are just about to cross Oxford Street, which in no way bears the élan of its name. Think Thirty-Fourth Street. Or Wilshire at Fairfax.

"I'm cold," I say.

"That's a warm coat," he says.

"I have to change."

"You don't like the coat."

That's true. I'd seen it reflected in some dark windows as we'd walked along. It looks awkward, but I say, "I'm just really cold." We go back and I change into the khaki leather coat from Loewe's. Excellent.

So we're coming down Bond Street. He's going past the windows as fast as the TV flicker, and here's Donna's store, all lit up, ready to open. And here's Donna—tall, wrapped in black wool up to her ears.

"See," I say, after we've hugged and looked around the store and gone on our way, "see, you never know who you'll run into. It's a good thing I changed."

"Donna's really going to remember that," he says, "as if you'd care what Debbie wears for Sunday lunch."

I think about all of us—Angie, Geraldine, Debbie, Svetlana, Judith and Laurie. I don't know what we wear. Only Mark dresses, when he can be there. But that's because he can't ever tell when Stuart will need him to drive somewhere.

We go to the café at the Intercontinental, where we will listen to an American jazz musician and drink Aqua Libra. They have a buffet where I can have sushi, which he won't eat, and he can have roast pork and everything else I won't eat.

I have worked out the longing to call Johanna back, to talk it over when it's too fresh will only make it harder. The thing to do is finish the book, and then go home to see my kids. Is home always where you started from?

Stuart's walking in his sturdy, rolling gait. I keep up with him easily, which is probably how he's worked out this walk. This street is more a mews: dark, winding, hidden, with doors open to narrow stairways lit by bare red lightbulbs. On the walls are pictures of women and terse job descriptions.

I look at Stuart in the dark lamplight. He once wrote, "Single men are designed around wall units, each woman another module."

Most men of his time spoke of women as numbers, car models, or graded them in numbers. Men did this especially when they were around other men, who assumed the more dismissive grief they said they gave women the more potent they appeared. Smart women (who believed this was a profitable approach) let men maintain that impression.

I hold his hand and listen to our footsteps along the cobblestone road.

26

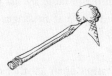

You don't have to watch me now, I want to say to Stuart as he leaves early this morning to meet with some business friends up north. I am okay. I know you'll come back. But then every time I hear your key in the door, or see you on the street coming to track me down, my heart skips a beat.

Just before he comes home, my friend and one of my favorite writers, Kennedy Frazier, calls me about a story for *Vanity Fair* that she doesn't want to do, a story she thought would be ideal for me. "It's sexy, Hollywood—not my kind of story, but perfect for you. So may they call you, is that all right?"

"Is that all right?" I say. "That's divine." Mainly what I hear is the chance to go to L.A., to see my kids. (As usual, I place them both conveniently in one spot.)

Five minutes after I hang up the phone I'm terrified. Where do I stay? Oxford is one thing. How will I manage L.A. on my own?

And then, to go home and not drive. But it will not happen. I won't get the job. That is the real dark fear.

I hear the key in the door. I'm not going to tell him about Kennedy's call.

"I had a bite after the meeting with Carl. Then Jeff joined us, all optimism of course."

"I figured."

He puts down his leather envelope on the hall table, flings his black raincoat over the banister, hurls the umbrella into the copper stand.

Stuart likes to do his own coffee. Then he goes into his gray room, tosses the evening paper onto the table by his blue-green leather chair. Then he sits at his desk and looks at messages and starts to open mail with his paper knife.

"Will Carl be all right?"

Stuart sips his coffee, slashes open another envelope, "It's a weakening proposition, scientifically and legally."

"How do you tell him?"

"You help him come up with a positive new proposal before you tell him how much trouble he's in."

"The sixteenth degree of denial."

"Of course," he says, "here I am trying to bolster his confidence about himself and his company when my own position is so vulnerable."

"You'll see." Our friend Mike has described us as "defying the laws of financial gravity." This is a particularly defiant time. "Something may come along." I could raise his hopes so fast, but this is no more than the kind of positive new proposal he invents for Carl. I stand behind him and rub his shoulders. "Don't do the mail now . . ."

I was going to say "like all men," but it's like all of us. He leaves so upbeat—crisp bow tie, fluffy hair. Then he comes back from a meeting with a company—any meeting, any company—telling me it went fine in a particularly deliberate way, which tells me nothing interesting happened and any expectations he might have invented didn't come off. This kind of reentry takes about twenty minutes.

I am glad I have not told him. No one has called. Would I really want to do the story? Maybe it's time I cleaned up my act. What act? Disappointment moves into heavy denial.

I can't leave him right now anyway. Next week Stuart is having angioplasty, a kind of surgery where they slip a slender wire into the heart to see how it feels, sometimes by that very act shocking it into action. Or that's how I see it. The doctor has tried to make it clear this is a routine procedure. We have made it serious, if not critical.

I am no more an accessory of his life than he is of mine; no more, no less, an accessory than the heart to the body.

I sleep in his arms the night before he goes to the hospital. Do I imagine his heartbeat is off? My grandmother lived outside the arena of the sexual century. I loved curling around her body, wrapped close in her arms. I'd press my head close to her breast, hearing her heart working hard to beat, praying these beats were okay. I used to try to count the beats.

I listen to Stuart's heart beating now. Does it flutter, like a moth trying to lift up when its wings are wet? Is this a long, drowsy beat, and has there been too long between this one and the one before?

We have talked about what happens if he dies. Before one of his operations, my father said to my godfather, "If I die, don't tell Miriam."

Stuart said, "When I die, our wills are upstairs in my dresser."

"Under the silk pocket handkerchiefs."

"Yes."

"With your latent supply of Cadbury's."

"Exactly. So you'll out my chocolates."

"You bet."

Judith comes with us to the hospital. I fill in the application blank, which has a place for "title." I put "Sir." They look at him and believe it.

Class behavior still works in a few corners of England; it works because the upper class is a security class. Upper-class people will provide things: work, a place to live, chocolates. It's a godparent class, handing down tidbits of reassurance.

Laurie is here waiting with me during the operation. "Don't clutch this hand," she tells me, "I'll never draw again. Sit on my other side. Here's my right hand."

The operation goes fine. This does not mean I don't need to watch him all this next night in the hospital, sleeping. The move of the triple circumflexes above his brows, now the right, now the left. They give sexy emphasis, a touch of surprise, where someone else might smile. His eyelids are shadowed in lavender and taupe, two of the better colors of the silk hankies in the drawer; heather like the Yorkshire moors, like the small veins in his big, smooth hands.

Over each brow there's a pearly knob underscored by the curvaceous frown line. The ram's horns were hacked off here at birth so no one would pick up that he is unreal.

After three hours, Laurie convinces me he will live even if I am not watching. She is clever. "Maybe *Vanity Fair* has called. You could, you know, tell me what the story is and you won't b'shri it."

"Maybe," I say.

I kiss his forehead, and we walk home together. I make some chicken soup, then bring some down to Laurie's flat.

Laurie's room is square with two arches. She is making color reproductions of medieval Biblical illustrations for a Dutch collector, and you almost expect to see some of those medieval figures standing in the arches. On the blue skirt of a woman doing laundry, "a symbolic washing away," Laurie explains, there's also the glow of firelight on her skirt. "See, I'm a middle-class artist with her feather duster," she says, as the gray and copper feathers whisk away microscopic pencil ash from the picture.

"Is the soup for us or really for Stuart?" Laurie asks.

"For him, but he's not eating. Making it is for me."

She has the chicken soup. I am lying on Laurie's couch while she listens to classical music with the particular silence of an artist working. This is like being with my mother in her studio—the gift of meditation. With her left hand lightly moving like a harpist, you can't tell whether it's music or drawing—the arts all mix together.

She sits in a sturdy black artist's chair, a kind of working chair that does everything; it adjusts for your feet, your back. On either side of her easel she has white carts on rollers with pencils, a really good straight-edge ruler, small dolls from a flea market—Ernest Borgnine, Bionic Woman, Cindy, Action Man, "much better than anatomical models," she's told me—music tapes divided by classical, pop and jazz ("my brother had his in alphabetical order, which frightened me, so I pulled them all out and had to put them all back") and racks of books.

"So tell me the story," she says.

"They want someone to find the girl everyone said Polanski raped. I could go to L.A."

"So what if she's not in L.A.?"

I had not considered this.

"She could be in Battersea, or Detroit," Laurie suggests.

"Where I have no family," I say.

"So," Laurie says, "what's to b'shri?"

Halfway through reading my pages that Sunday, a riff about my father and Adlai Stevenson, I stop and I tell the writers about the *Vanity Fair* proposition.

"You're packing now," Mark says. They're ecstatic, encouraging.

"Stuart will never trust me to go by myself," I say.

Oxford was miles. This is time travel. I could go and never come back to the present.

"Do you trust yourself?" Geraldine asks. "That's the point."

"So . . . ," I'm careful as I tell Stuart about the story that night, "I haven't been to L.A. for years. But Jeremy does want me to be with Phoebe. I haven't asked, but I'm sure I could stay with him."

"We were there last year, Jill. You wrote about L.A. for a travel magazine. I played the piano, Phoebe danced, and you had a family reunion."

"Don't say it like an accusation. You have reunions, too. And family."

"Family," he says, looking at mail, "nostalgia . . . melodrama . . . and severe editing."

Thinking about my family leads to considerations of denial. Losing memory is sometimes yesterday's denial—and I do that. Denial is not acknowledging what I see right now. Then, when I do find it in the memory store later, it will have been redone, will look the way I want it to look.

I see it in the way I can handle it, although I am trying, here, not only to remember, but to remember that time as it was. But doesn't that mean as I saw it? Aren't I accused, if lovingly, of seeing things in the sentimental spectrum of old Technicolor?

It's just that I don't think reality is always the harsh side; you don't always have to look with the cold, clear eye. It's an irony that the sexual era coexists with this looking at everything real. Sex does so well with romance, which needs to wear a little denial behind the ears, to flick it on the lashes and spray it on the throat before you say a word.

"I've almost finished reading your journals," I say to Stuart this morning. "I stayed up all night."

As the journals progress, they're filled with more scenes of his own memory, tough writing about the years at Procter & Gamble, the part of his life I don't want to see, when he seemed to be the contained forceful male, like my father was when he ran MGM. In his journals, Stuart writes about the fear, the restlessness, the conflicts, and how they're brought into focus by dealing with me, caring for me. Once again, he is the contained male. The story he wants to write about his own life is jetting out of these journals.

"Were the journals helpful?" he asks quietly. He's reading the *Herald Tribune* and I'm distracting both of us from the issue at hand—L.A. on my own.

I'm kneeling by his blue-green armchair, stroking his pink flannel robe, the one with pictures of ice cream cones on it. "I never saw the pain and difficulty you went through, not just because of how little I registered during these years, but because of my fragility—you've been so trapped. You must have felt like a keeper stopping me from being lost, confused, or frightened into blowing a fuse forever. You also dealt with such tempers!"

"Those were not new," he says wryly, "but you did throw yourself into the street in front of cars more often. Your temper was just exaggerated by the early medications." There were scenes he had to just mellow down for fear I'd work myself up into a seizure. And there was always the question, was I aware he wouldn't retaliate because he was scared I'd crack?

"Stuart, listen, through every year," I continue, "as you write in your journals, you talk about not writing your book. Yet you're writing it. It's here, in the journals. You never told me how much you were missing it, how hard it must have been doing work you hated so that I could lie around trying to connect fragments of sentences."

"I will write my book when the time is revealed." He already looks tired of the conflict he'd have to have with himself to write it.

"In 1990," I remind him, "you wrote, 'I won't be writing in 1990,' and you also wrote, 'That is unacceptable.'

"You also said that some consultants concentrated on making money—others, you, as you said, 'tried to perpetuate a dream of management.' You even try to make business creative, like a figurehead of a character in an old skyscraper movie. Stuart, you believe so in whatever you do. You didn't think you could support your family by writing, so you thought of business like writing or directing, having an image, working with people, shaping their stories, like a director. It's also like teaching, like counseling, which I've watched you do brilliantly. Maybe that is what you want."

He hates when I do this. "I can't talk about this now. I need to get dressed, take a walk to pick up some coffee."

"Of course, the minute I start to talk about you—your writing—or not writing—you have something you need to go and pick up."

"Jill," he stops, "I watch you. I've seen how it was. I'm the only one who knows." He switches glasses, puts down the newspaper. "I can't pretend it would be easy to see you go."

He sits forward in the chair. "I can't ignore what I see, even when I don't tell you."

"What you think you see is your mom."

"I can't help that. I see how hopeful it was when the new doctor changed her medication. Our phone calls were lively, encouraging. She kept up the snappy exchanges that lure me into calling. She reacted with excitement, not fear, to the notion of visiting London, the way, on good days, you have no trouble thinking about going anywhere. But a day later and she's gone again. She's forgotten all the joking, the teasing, the poems, the organ playing, the old songs, histories retold, days, nights, and old times recast, reshaped to fit the present; the Christmas hams, the Stilton cheeses, the chocolate Yule logs."

"You think I'll go up and down like your mother, and wind up out there down and alone. But I'm not her. I don't have what she has. And I'm not alone. My kids do care."

"But,"—it's heating up—"right here—look at right now—you've forgotten that your children and I do get along. I'm the one who reminds you. You forgot Jeremy's Thanksgiving visit when Phoebe was still in their arms. You forgot how he strolled with me and we talked of buying land together."

"But when you remind me of things—you remind me of your mother."

"Stuart, I have my weird days—but hardly as many. I know you're afraid of me going somewhere. Maybe that's what's happening. But it will never happen," I tell Stuart, "because they'll find out I can't write anymore."

"Look, they know you've written several books. Has that slipped your mind?"

"That was before. And the last one got reviewed in New Jersey."

"The condition was to write, to be published—and you've done that."

"You're cranky," I say, "because you don't want me to go."

"No," he says, "you're cranky because you're scared to go, and you think if I get mad, I'll stop you. I don't want that. This is a great chance for you."

C·H·A·P·T·E·R

27

*S*ex was the opening move in L.A. conversation in the sixties. Now it's "What's your favorite meeting?" or "Where does your kid go to school?"

It's very likely, with the sensibility I have now, that I choose to see those days as evil. Most of the books and articles I read, most of the people I talk to, speak from today's sober approach. In the sixties, we spent more energy looking back in anger at our parents than we put into our own kids. Our children have become the parents we now wish we had been.

First of all, I want to talk to Polanski. I find someone who will give him my number. I call people who will be interesting about that time; people who might have clues to the girl; people who knew Polanski; people who were there.

For most of us, far more important was the murder of Polanski's wife, a bloody enough event to throw any man. So what, I think when I read the two stories again with today's attitude. We threw him out of America for this?

I call Rafelson and his ex-wife, Toby. I call Brooke Hayward and Dennis Hopper. Brooke says, "Polanski is a bitter man." She last saw Polanski in Paris in 1981. "He hadn't felt he'd made a great film in years." That was long ago the way lives go these days.

And then you can talk to any artist on any given day, or night, and hear that all the work they've done sucks rats. "Don't trust any artist who doesn't doubt his work," my mother said.

I decide to call someone else. Someone who was, as I saw it, outside the sixties scene, I think.

I'm not sure Bob Redford will remember, but we were in class together at school. I like the pictures he chooses to make. And he keeps a balanced perspective. He's involved and maintains a cool, objective eye at the same time.

His assistant was careful. She'd get back to me, she said.

Gillian Farrell, the mystery writer and Larry Beinhart's wife, reminds me, "Polanski's name came up a year ago when we were here for dinner. It was just after you did that travel story on L.A."

"I did?"

Andrea Tana, the painter with the auburn hair, was here that night. They all come over again to refresh my memory. "We agreed that from the Hollywood perspective, the sixties ended when Manson killed Sharon Tate in 1969," Andrea says. "I heard the people who bought that house have tours through it—like a monument to murder."

The candles on the table tonight flicker. In even the worst movie of this scene, they would not do that.

"Is Polanski still in Paris?" Larry asks. Larry's rugged John Garfield face is directed by challenge.

"Sure," Gillian says, "in exile. Remember we wondered what happened to her?"

"I still think it was just a story—someone set him up. If it was real she'd have been on TV, written a best-seller by now. No one even knew who it was."

By "no one" we mean us, privileged insiders. If we don't know, it isn't true—never happened.

"But it was this young actress," Andrea mentions her name, "I know that."

"I've never heard of her," Larry says.

"Young actress," Gillian says, "not a big star."

"Clearly," I say, "since no one has her name right."

"But why wouldn't she talk about it?"

"Someone she loves doesn't want her to. She's in her thirties now. Would you want to talk about it if you had kids?"

"Do we know if she has kids?"

"TV actress in L.A. doing nothing. Why hasn't she come out with it?" Sure. Why wouldn't she want the bestseller, the movie deal? Suppositions work the room. Gillian says she'll talk to her friend Leuci, a great detective in Rhode Island.

She calls the next evening. "He'll talk to you."

Leuci has the voice. You'd cast him right from "Hello."

"Where do you begin?"

He says he has an idea, and he'll call me back. Each phone call has that waiting time, like waiting for a date.

"The public never forgave Roman because someone murdered his wife and shocked them," Toby Rafelson says. She has a salty, wise Eastern voice, Sara Mankiewicz. The Rafelsons' house was a place to talk, to eat. L.A.'s Central Park West.

"He was supposed to be chaste forever. Perhaps Roman's exile is an anxious xenophobic reaction to his dark work, a confusion of what happened to Sharon. Somehow the public believed if he had never been there this would never have happened."

I want to hear from my agent and from the detective, who are both doing the same job, looking for something I don't believe I'll get. Who will call back first?

Waiting. Writers do this very well.

The trick is to get out of the house before you think the phone might ring. Then you do not watch the phone. And when it is a call from New York you are waiting for, it will not come until you sit down to eat dinner. The phone is far more likely to ring, however, if you are out.

But the call will not come on Monday. People are then dealing with stuff they didn't do on Friday, and they've come in late because of the weekend traffic.

Tuesday they are dealing with urgent things. Writers are not urgent.

Wednesday is their best day to do lunch. After a long lunch, there's barely time to go back to the office before you have to meet someone over drinks. (Meeting over drinks is work talk. Meeting for drinks is another thing.)

After drinks, it will be too late to call London.

Thursday could be a day for calls. But then they're cleaning up for Friday, when it's just as well to work at home and miss the weekend traffic.

I contemplate the story. I reread Cheryl Crane's tough book about her shattering childhood and triumphant recovery. Raped and battered by two of her mother's guys, accused of murder when she was trying to save her mother's life, Cheryl was raped from the time she was ten. This might give me a clue to this girl's angle now.

Cheryl's mom, Lana Turner, was the perfect silver screen star in black and white, all creamy, pearly luster; complete glamour, and marrying, marrying, marrying. Cheryl used to watch movies in an empty screening room. Bob Topping told Cheryl her real father, Stephen Crane, a nightclub owner, had died, so Cheryl would call Bob "Dad." Loretta Young won't admit, even now, that Judy is her own child. Judy thought she was adopted until her wedding night, when Jack Haley Jr., son of the Tin Man in *The Wizard of Oz*, told her she really looked a lot like her father, Clark Gable. My mother told me during one of the long night talks just after she'd finished Judy's portrait.

"She looks more like her father there," I said.

"I think so," my mother said.

I asked Judy if she ever saw her father, and this was long ago, when she said only once, when she was sixteen and getting out of the pool in her new red bathing suit. And there was Gable at the garden gate, looking.

"Perfect," I said and knew I'd steal it for a story.

Big tanned Tarzan Lex Barker took Cheryl Crane into the sauna, where she'd never been. Where the only light came from the dark gold bulb on the ceiling; quizzing her, touching her, "did he ever do anything like this?" He took it out, rubbed it, said it was his rabbit. "Remember, from now on this is going to be our secret, ya got that?"

You see a spirit ennobled by a lousy childhood; a girl put into reform school for saving her mother's life by stabbing her lover. Naive, scared, she slashed him sharp side up. Cheryl was her mom's protector. The giant child-like ego needs all the attention it can get. The wise child attempts to get anything from its parent, so it becomes the listener, the home audience, the comforter, the parent.

Jeremy and Johanna could dine out well with Cheryl.

Gillian calls me a week later, after midnight. She's excited. "Leuci's trying to reach you, and he has someone for you to talk to."

Leuci's daughter, Santina, produces the show called *Hard Copy*. Larry Silver, this lawyer for the girl who was raped, has called her. He wants $100,000 for the story.

When I actually get Silver on the phone, I tell him, "I want to do something serious, to show how kids can reinvent their lives." I talk about how the fame plague rips right down through families, about the marketing of kids from the beginning of time.

I talk fast, tell Silver I've been there, written about it, "every woman who's been drugged or drunk knows about rape. It's not something you want to explore again if you get through it."

He's listening. He's got hearings, he'll make some space if I call him next week.

"I know I shouldn't have trusted Santina," Silver says. He's had a "week from hell." He's going to try to turn it into an epic. Some people who do that make it worse. But then, some come through better than the ones who say, "Sure, right away!"

"It's *Vanity Fair*," Stuart says.

That's early. I dash to the phone. It's all over, probably. There's no hello, and I never heard of this person. But then, if I had, I wouldn't remember.

"We had a call from Robert Redford's office." This is not my editor's voice, not George. This voice has attitude. It's in the job description. "We don't like our press to call stars unless you've been given clearance."

"I am not press!" I say, shaking. Anger, embarrassment, or fear. Who cares. Let it sound imperious. Think Hepburn. "Look, I know movie stars you haven't even heard of. Redford and I went to school together long ago when Sepulveda was just a pass and there were trolleys downtown. Don't tell me anything about L.A., its people, or who I can talk to," and hang up the phone.

"Terrific!" Stuart says. "You did that well."

"And killed the job. And I just vaguely remember Redford. He won't remember me."

I spread the Liberty cloths with the big scarlet poppies across the dining room table. One cloth has an ivory background, the other black. The poppies remind me of the acres of poinsettia fields stretching from back of the Sawtelle Veterans Administration Hospital, all the way along the Sepulveda Pass, back when that was the main way to the Valley.

I do know L.A. That is, and was, my town. This kid assistant wouldn't know what I mean if I talk about the poinsettia fields that carried on about as far as the tunnel.

Debbie, as always, is the first of the writers to arrive. She dresses like a very small, impoverished French model, with tiny tops and skirts that no one else could figure out how to wear. Her black curls mix into the shaggy black scarf. She hands me tulips and begins untangling herself from her coat, shoulder bags, and sweaters.

"I was thinking of you and all the men you must have slept with," Geraldine says, "if I can say that."

"Well, you just did," says Debbie.

"But I mean," Geraldine says, "how hard it must be to have forgotten all that. You can study us for women characters, but who can you turn to when you want to write a man's character who isn't Stuart?"

"No one says she can't look," Judith says.

Do I even bother to remember? Or are they all Stuart, which is why it's not working?

"Can I read," Judith says, "because I have to leave early."

So she reads, "Sylvia wrote a list of what men do. She rode the tube. She watched them for a whole day: jeans hitchers, belt hikers, watch checkers, wallet tappers, ear scratchers, brow twirlers, neck stretchers, shruggers, hand wranglers, chin rubbers with their invisible beard, or no beard at all, but pulling jaw skin up, nose rubbers and pickers, hand washers and crackers, ball shifters, bald guys rubbing like that would bring the hair back. And then guys with glasses got a whole world: pulling them down to rub their noses, looking up over them, taking them off, twirling, spinning, clicking.

"It's hard to find one," she reads, twirling her pencil between her right forefinger and thumb, jiggling her right knee and twitching her left ankle, "hard to find one who sits still, and harder than that to find one you'd want to sleep with."

"Well," Debbie says, "you do look." Debbie is still slight, translucent with postadolescence, which makes her sharp sentences startling.

Judith glares, "But it's good, isn't it?"

"Great," Angie says, "you can just hear Sylvia!"

"I like that. You're so good when you keep her close to you. You watch them like you're making a movie."

"Because that's what Sylvia's character wants to do," Judith snaps.

After Judith leaves, we have some tea and honey and eat the muffins.

"So," I say, "I was reading about the Acropolis in the memory palace book. It was the most important image of the ancient Athenian world, its crowning image—and I was wondering if each story or book has a crowning image, a moment we're going for that a particular character never loses sight of. Finding the girl is as far as I can imagine now."

Can I imagine that far? Does memory's break amputate the ability to project, to imagine ahead? My Acropolis today is really seeing L.A.—getting home.

Debbie is also almost always the last to leave. I'm in the kitchen when the phone rings. Debbie's bringing in the ice cream dishes, which now only hold cut-up fruit. Have our writing styles changed as much as our eating habits?

"Would you pick that up, luv?" You learn to say "love," not "darling," and after five years, you say it with some slight shift, which means it's got a "u" in the middle. (Don't try. You have to be here.)

"Jill," Debbie comes whispering in, her cheeks scarlet, "he said, 'Hi, this is Robert Redford.' And it is!"

"Okay, okay." I fluff my hair in the mirror over the kitchen sink and dash to the phone.

"Hi," I say, "I'm sorry. I didn't mean to bother you, but I think you know me . . ."

"Jill, you think I'll know you?!? We've known each other since school and I was in one of your dad's plays, don't you remember? It was my first try on Broadway."

I explain to him about the memory thing, leaving out the epilepsy bit. I also hate to say it was a stroke, which sounds old, so I probably said what I do, which is "this thing happened and I was in a coma and lost my memory." No one asks "what thing?" if you say it all fast enough.

"I'd love to see you and talk to you. Just call when you get to L.A."

"So, you believe in keeping everything you have ever worn?" Valerie Wade has come to help me pack for L.A. I have not let too many friends into my changing room. It is here that I become who I will be today.

"Now I understand how you do it," she says, "you make up outfits the way I make up rooms." Valerie has the definitive interior design shop on King's Road. "You take old pieces and invent another look for them. But what could this have been?"

She holds up the patio circle skirt Loretta Young made for my mother the Christmas "everyone was doing one." It has royal blue and emerald green hydrangeas and is scattered with sequins.

"It's wonderful with a chartreuse ascot shirt and a delphinium cummerbund."

"I don't see it for slipping in and out of rooms trying to find the girl."

"I may have found her." I put the skirt aside. And Ralphie's ancient knitted golden cardigan.

"You're kidding! Not really?!"

"One of the Fulbright award winners, you know, for detective novels, knows this lawyer in Boston whom I called. The lawyer sounds exactly like the guy in *NYPD Blue*—Slovensky or something." The man you marry after the one you knew you shouldn't have married gets away. Cranky. Smart. Heavy. Maybe Slivowitz, protective and loyal. "Anyway, he says he thinks he may have an idea but he'll have to make a call. It may be nothing."

I toss the gold sweater into the bag. "There is nothing better than this and a white shirt."

"Except the tan Donna Karan one you've already packed. But I mustn't discourage you. So, have you heard back from Polanski?"

"Nothing. He told me to leave him alone. I don't want to find her before I get there anyway."

"This is something you tell us all the time."

"So, it's no news?"

"None."

"I'm not sure Jeremy remembers me."

"Are you sure you remember Jeremy?"

"Fair question." Tough question.

We wanted to raise our children in a way different from the way our own parents did. I remember bedtime was of no interest, and why go to school if a peace march was coming together. This was political consciousness, just a new take on my father's approach.

The night before I go, Polanski does call me back from Paris. He's charming. I don't want to lie to him, but I try to talk to him about the changes in Hollywood. He was taken in, he tells me, by the "ingenuous time of the sixties in Hollywood"—the notion that everything was all right, that you could do anything.

He says, "The sixties ended when Kennedy was killed. Sharon's murder and the landing on the moon changed everything—when man walked on the moon some romantic idea of the moon was over. I believed in romance when I was with her—but then the magic illusion was gone."

We talk about the fear, the bitterness running through his dark, complex love stories. Could innocence still turn him on?

He's careful. "The artist catches violence. If it's there in the culture, he'll reflect it."

I am thinking of questions, such as, "Did your own longing freeze at that time? Did you need someone new to begin again? Would you have bolted if you did it again now?" These are questions I might better ask myself.

"But," Polanski says, "we all did things we might regret. Sex and power and celebrity are all the same in America—which has taken the mystery out of sex." He spoke dismissively. "Now when you speak to an ingenue, you hear a baby crying in the background." Like many great directors, he's swift and icy, until you're into the scene he wants and then he's totally there with you, until he wants to move on.

"If you read my autobiography you will see all you need to know about that day."

He says he hasn't time to see me. And there's no point in telling him his autobiography is out of print and not in London libraries.

I saw my brain in a dream last night, spread out around me like this oatmeal cashmere blanket Jeremy sent me. I was stroking it to appreciate the things it was doing, showing it off in fact, when I saw under a puff of the blanket there was a fire. I put it out quickly only to see another, and on and on, as if all over neurons were exploding into tiny bonfires.

Do I trust you, brain, from one day to the next? Can I go along to this corner, certain you're not sabotaging another place I've forgotten to lay my attention? Can I catch it all?

29

I am flying home, jetting back to a world I know.

Last night I reread bits of *The Cause,* the novel I wrote about the sixties for Ed Doctorow when he was a young editor. I would chronicle the world I felt bursting around me. This was my destiny. I was full of naive drive—and speed. I couldn't make the last draft.

This story is different. I know about rape. I know the sixties. I am this girl.

Stuart's son Philip calls it "The Golden Age of Sex." "So, what was it like?" he said when he visited. He's blond, with loose, long-limbed, an-idea-a-minute vitality. If there are period faces, Philip has the tough, tender thirties man face. Maybe it was last week—or Thanksgiving. The point is what he meant was the time after the Pill and before AIDS.

I was just as fast then, except I'd take stuff so I could listen more intently to my mentors and coaches. I must be something to be in their awesome presence. I never saw their desperation. They needed my young wonder as much as I needed their wisdom.

At thirty I almost killed myself because Sylvia Plath was a famous poet and I wasn't. What chance would life have if you weren't a legend by thirty?

It isn't the custom now to speak of sex in our old easy way, in the style of the "Golden Age," as Philip put it. Sex was a craft then, as readily picked up and discussed as basket weaving. To keep its mystique alive, we had to come up with new ideas, new taboos to spring free of. Initiative and imagination mattered.

When you're young you think you can get away with anything—now that I'm grown, I see what a hustler I was. I no longer wanted to be seen as this kid who had everything. To be a real writer I needed to know what real trouble was like. If I'd held on, life would have come to me. But I went looking. I wanted to walk the high wire.

My first marriage, to my children's father, was crumbling. I was hardly trying. My book came first, neck and neck with sex. This was 1962 and the art of sex was in the pretense, in the invention of new ways to lie together. "We did it standing!" I whispered to one of my girlfriends who had been married longer. "So," she shrugged, "we did it with the lights on."

"And your eyes open?"

"Really," she nodded.

No one talks now about how it felt then to suspect there was something more to sex, something more than even these gymnastic positions we thought we'd discovered.

My first husband looked like Raymond Burr, who played a detective in a TV series; every description was in reference to a TV or movie star, or a car. He also looked like a mid-fifties Buick. This was a heavy car with large swashes of chrome. Escada clothes look like Buicks, so in a way my first husband reminded me of a line of clothes that hadn't been invented then. His mother, on her fifth marriage at the time, would have been comfortable in Escada.

On this night he had fallen asleep again in front of the TV. This made the room the color of a black-and-white movie. He had his cigarettes in the pocket of his pajama shirt. He was snoring softly, but looked entirely attractive in the slate-gray light.

I felt damp and raw and breathless with longing, and I'd taken off my clothes and poured myself over him the way you do, kissing, licking, reaching. He was heavy on me; I was a treadmill, a trampoline, and the thrill of just the idea of doing sex had gone. "Doesn't anything satisfy you?" he said. There had to be a catch to it, and I didn't know what it was.

I wanted it to feel at least as surreal as the sensations I'd have a few times during the day. I thought they were from speed.

They felt like falling and flying all at once—like dreams I'd had of how sex would be. So I went to see a serious Freudian analyst who had experienced enough, we all believed, to tell us hard things. He had known all our parents. He'd look at you at family gatherings and you'd dig in for your worst ideas and figure, if you were feeling lust, envy, or fear, you were onto an authentic track. He liked to use Latin sex words around kids. Our parents snapped to attention. We'd blush; he'd wink. I'd read it all in Latin, the only foreign language I learned carefully because the very good parts of Havelock Ellis's *The Psychology of Sex* were all in Latin. I didn't know exactly what they meant, but I'd whisper "fellatio" and "cunnilingus" to myself when I thought of Gable's leer or Brando's loins.

"Many men don't understand women's sexual response," the analyst said. The words were unbearably hot. Did he want to show me?

I explained to my husband that there must be another way to make it work. But every time I'd get close, I'd feel the way I did as a child, when naked, raw images crackled in my head with a warm hum, the world would go black, and I'd wake up somewhere else. If this was coming, I couldn't risk getting to the other side.

We were trying once again. Hating the surreal images creeping into my brain, I wanted it to be all his fault. I wanted to stop, but you couldn't be the quitter. It would be a long time before imaginative coaches got me to take the leap, convincing me, even, that the suspense was seductive.

The phone rang just in time. "What?" I shouted into the receiver.

"Bobbie's in trouble," I snapped at my husband. "I have to go."

"That's one more excuse to go hang around with all your friends." This was, of course, true.

Bobbie, my friend Roberta Neiman, was not in trouble, but she could hear I was. Bobbie said she'd pick me up on her way to the Ferus gallery.

Every Monday night the art galleries on La Cienega were open late. The artists Ed Ruscha, Ed Moss, and Andy Warhol and dealers like the Easterner Richard Feigen, who visited every summer since he was a teenager, and Molly Barnes, who had always been part of our world, were here to see the scene. Artists cruised, checking out the space, the crowds, and the prices the dealer put on their buddies' work. Later, after the dinners at El Coyote, much later at the Beanery, they'd talk about the work, their dreams of glory.

On most Mondays the time we spent hanging out after the galleries closed began to last longer than the time looking at the pictures. We'd start at one gallery and wander on. Unless there was an opening, like tonight. And the pictures, like Ed Kienholz's assemblages, were sometimes not pictures. Art, like our lives, sprawled and leapt outside its frames. The art spilled out first, then we followed. The things Kienholz assembled for us to stroll around were pictures from the darker images of *True Confessions*.

Larry Bell and James Rosenquist (and, when they were around, Roy Lichtenstein and David Hockney) used to come over to Lynn and Don Factors' on Sundays. Theirs might have been the first severely modern house on a Beverly Hills street, stark and white in a procession of Venetian, Moroccan, Federal, and Tudor mansions. Hockney wore tennis flannels and so did Irving Blum. Feigen wore ivy league shirts. Don Factor had a court, so maybe he used to play. When my father broke his back and couldn't play anymore, he tore down the tennis court.

"You know," Lynn Factor said, standing there in her smooth new JAX dress, "these are like the tableaux we used to have around Christmastime along Little Santa Monica, little stables full of straw. It's like you've done dark tableaux."

"I've been waiting for you to tell me what I've done," Kienholz says, sucking at his cigar and looking her over. Bald and burly like a Harley rider, his chin's a match for his beer belly. Tiny, piercing eyes. I thought of Sylvia Plath's poem about the Nazi. How dark can I go, how down? Can I file away the elegant cover of my family image? Can I show someone who I really am, and then see, somehow, if I can be loved even so? Can I be someone even so? Can you touch me now? What about this? And after this and look at this—not now, you'll never look back now—testing, testing, how far can I go down and still get back?

I'd learned that the way to pass for hip was to say nothing.

He looked me up and down and asked if I'd like a joint. I didn't like pot so I sidestepped, asked him to be on my radio show.

"You don't want me on your show," Kienholz said.

He looked like a construction worker with his jeans down low, backside cleavage and burly shoulders. But he did do his constructions after learning the academic way. Like my mother said, you didn't take off into new abstract worlds until you knew what you were taking off from, until your own line was steady and clear enough to give your work a presence.

We were standing in front of his royal blue car assemblage with the couple doing heavy petting in the backseat. "I'm not about talking 'look at what I do,'" he said.

"Maybe you'll see what you really want to think about," I swung around to the door. "I have a deadline I have to make."

"Make me," he said so quietly.

"Later," I said and fled.

I had an office by then in the Writers' Building in Beverly Hills. The building, long and not too high, was built to last through a good quake. My office was on the third floor. I found a bright yellow Mexican rag rug and paper flowers on Olvera Street one Sunday when I took my kids downtown for tacos. The office was twenty-five dollars a month and my payments were usually late.

I said I took the office so I would discipline my writing. I'd go for several hours every day. I couldn't work at home surrounded by reverence in every antique bank, every painting and silver candelabra; chaos reflected in every unanswered letter, crumpled manuscript, and smashed glass, leftovers from last night's fight, which I'd triggered to distract from the terror of writing.

I had a typewriter in my office, packages of typing paper, and yellow pencils. I knew not to have a phone. The screenwriter downstairs, married to my son's godmother, let me give the number to Nannie.

Here I'd try to finish the novel about the woman looking for a cause to give her life to, as her liberal father and his friends seemed to do during the war and the years of blacklisting. I had no such gift of focus, no real fervor in spite of the snappy jargon I tossed around on TV with Mort Sahl. The only thing that drove me forward, screaming along empty beaches, was sex.

But you couldn't write about that.

I had my father's old writing desk in my office and a canvas director's chair. "I need something to lie down on," I told Brooke and Dennis. "You have to sleep yourself into some scenes and stories."

"Kienholz has an old psychiatrist's couch that would be ideal," Dennis said.

"Just think," Brooke said, "of the ghosts of old Freudian dreams you can pick up."

I met up with Kienholz at Barney's and followed him up to his house in my car. You could make it with any artist (even if you weren't so sure it was art, after all), but you couldn't drive in a car with any man if you had kids you had to support.

Some of this was the beginning of feminism. For art, it was the last decade of the presumed male.

His pickup truck sputtered and chugged up the canyon, and I staggered behind him on the trails to his house, where he climbed over his harem of embracing, lunging, hovering mannequins; bodies and limbs like car parts (some were car parts).

"She can't collect art," Lynn Factor once told a photographer from New York she introduced me to, "so she's collecting artists." She laughed.

"In what way?" Her friend leered down over me, his big square jaw shadowing his ascot printed with peacock plumes.

"I eat them for breakfast."

"No wonder art in L.A. is so prolific. Heady meals. I knew it," He said.

Kienholz's territory was like a hillbilly hideout ambling round in the far back of Laurel Canyon. There were lizards drowsing on the old car parts, a large one with blue flecked paint on its stubbly tail, and I knew he'd look like that, stubby, when he slipped his jeans down even farther. The nakedest part of a man anyway is the back of his neck, where you'd pick him up with your jaw and carry him off to your cave. You own him when you bite him there.

The lump on his neck rose above his undershirt. There was more than one lump, a varied musculature of severe power. I thought it best to back off. But I did him. You try to think of it as conferring a kind of artistic grant.

You could look down at night and see La Cienega, which he said was as close as he wanted to be to the art scene. No artist says he wants to be part of the scene, but in the beginning they have to play it to survive. Hang out in a corner, look angry, depressed, or chilly. Come late. Leave early. Do not dress in any hip way, or like an average artist in jeans and a shirt.

My art teacher, Howard Warshaw, gave an impact to the wombat in his pictures that you wouldn't have suspected, and laid more truth on the rule that the worst-dressed guy at the gallery was the best artist—until art got hot and artists wore Ralph Lauren. I am not talking about women artists, for this was the edge of the sixties and there were no women artists to speak of in L.A.

If there were, I didn't want to hear about them because my mother was as good as any of them—and I didn't want to know how she was feeling. You don't know some things until it's too late.

I didn't kiss Kienholz that day. It wouldn't exactly be adultery (still a major concept of some deep concern back then) if you didn't do certain things.

You would not kiss lips.

You would not stroke hair. No problem with Kienholz, he had none to think about.

You would not say love words such as "darling."

You would not smile up at them (or down at them, depending upon where you happened to be in the arrangement).

You would keep it tough.

"Snakes are aphrodisiacs to German shepherds," Kienholz pointed out as we finished this turn through his chaparral trailer park territory, winding through tossed-off ideas, sawed-off mannequins, loops of cloth, heaps of limbs, tins, paints, stone, marble, board, canvas, and twine.

Artists like to tell you scientific facts so you don't think they have their heads up there all day in the clouds.

I could see that his German shepherd was taking his time over a rattlesnake he was just finishing up. I turned toward Ed. "This is like the kind of question I'll ask you." (I was throwing "like" and "sort of" into sentences to give the impression I had been an early beat person from the North.) "What makes you think you're a real artist? When do you know?"

"You don't ask the question if you are," he said, rubbing his forefinger along my backbone. I had my clothes off by now. This was no longer a strange phenomenon for me: neither having them off with a stranger, nor suddenly discovering they were off when only a second before they had seemed to be on. These discoveries were one of the amusements of the new chemistry of the time.

The dog had come along into the room where I was sitting on the black leather chaise longue. "And you don't talk all the fucking time about art in our time and all that bullshit, and you don't say no to anything. Schultz wants you." Schultz is the German shepherd.

I could do a lot of dangerous things in one go. "Schultz is German, uncircumcised, and has rattlesnake stuff in his mouth," I pointed out. "I am allergic to dogs and he's turned on by snake venom. So don't try to tell me the dog thinks I'm someone."

I am someone, but my definition and Schultz's definition of "someone" are far removed.

I started writing a screenplay. It would not be about marriage. You could not have a good marriage and be a real writer. Choose one. I chose. It seemed natural to make it a movie. I could see the picture. It would be the story of a woman who was a stylish wife and mother by day, and a man, an artist, at night.

I watched Brooke and Dennis. Brooke was everything the woman should be, elegant and clinically aware. Dennis was everything the man must be, romantically sullen, difficult, radical.

One actress would play both parts. But when she played Jordan, the man, an actor's voice would be dubbed in. The sets would be designed by the new artists. The crazier I got, the darker the book's ending became. At first the woman killed the male artist. She burned his pictures and freed herself to take care of her children.

I heated up my drive to write by talking around pools and in living rooms with friends. Jeremy and Johanna were friends with Peter and Julie Rafelson. Their parents, Bob and Toby, landed in the center of our lives with East Coast authority. I watched Toby host the mixture of artists, filmmakers, and Easterners who would come out to observe Hollywood's transition, and I wondered what happened to us. There was no question when I'd sit on the high backseats in our projection room watching friends that Warner LeRoy would be running the town. If I could have chosen the son to be, it would have been Warner.

I'd assumed, growing up, that we'd slip into our parents' or even grandparents' roles with easy aristocratic grace. Assumption doesn't do the work. Charm doesn't help. It only makes you feel even more entitled. While you're lounging back telling entrancing stories, the ones you aren't inviting are at home, working out how to make the goals you say you have happen.

Us. Who were we? Usually whomever Josie Mankiewicz had sitting nearest to her at the Sand and Sea Club. Even Sharon Disney came down from her club to be near Josie. Nora Ephron was different. Her mother was a working writer, which may have been one of the most useful things that could happen to determination and wit. You don't know how different the attitude toward working women was then.

It was another thing to have a mother who was a star. This was an automatic tragedy, but in those days that didn't mean a book deal. It meant denial.

I thought I wrote the scene about the woman in the book burning the man's (her own) pictures the night after Brooke called me, sobbing, to say that Dennis was going to kill her, could she come with the kids and stay with us. For $135 a month, I rented a wonderful small house overlooking the sea in Santa Monica Canyon, with a living room like an artist's studio and more than enough space.

"He put his paintings out on the lawn to dry—and I turned on the sprinklers. We'd just planted a new lawn after the fire and I'm very careful about it. It's been so dry."

Lawns are a source of accomplishment and trouble in L.A. When I was around eleven or so, I was furious with Angelo, our gardener, for telling my father that these boys I had over, Bob Redford and Billy Keane, were kicking holes in our lawn playing football. My father was working at home. He came out, yelled at them, and sent them home. I was embarrassed. Devastated.

I'd actually invited only Bob because he was good at drawing and I thought he'd be interested in my mother's studio. I'd been disappointed when he brought Bill, but boys did that then. You never did get to see one on his own. If you really got along, then what could you do about it? You couldn't just be friends; someone for sure would point out that this was strange. What kind of guy has a girl as a friend?

I was friends with Dennis. He'd been on my radio show. We talked, the way I talked with the other artists, about the work and the impact the change in art was having on Hollywood, on behavior and attitude. Was it, I wondered, the other way around? You could talk that over for a long time.

Dennis would show up at least once or twice a week, and I'd read scenes. I wanted Jordan, the artist in my screenplay, to sound like Dennis, the artist wrestling with fame and his talent's hundred or so hot drives.

Dennis listened. "You're way off." I'd try again. "I feel so different and scared. Like I don't know what I'm doing here and I don't want to fuck it up," he'd flash this grin, "but I am, aren't I?"

"Absolutely."

Looking back, I know I was in love with the idea that I could share this drive, this art, with a man. We could be wild, creative, on the same wavelength, all new. It was a revolution.

I was never this girl I am looking for now.

Sunset. I see it from the air as we turn in for the landing. That's all I need to see to know I'm here. Phoebe is here to meet me, with Jeremy, who has his grandfather's smile; the rest is Bruce Willis. He has what my father used to call "the motor." I know where he is standing before I am off the plane.

I am at Jeremy's house only five minutes before Judith calls from London. "My mother knows a young guy who would be interested in driving you around."

"Terrific," I say. "So, how are you doing?" I ask. I pull her face into focus. Yes. It's here.

"You've hardly left," she says. "I'm fine."

I'm watching Jeremy on his other phone—foot jiggling, clicking channels, and reading. It occurs to me how much Jeremy and Judith may be alike. Above all, they are desperate not to let anyone know how much they do for people they care about. Judith's seen to it that I have a driver. There's a dinner party tonight and my luggage is on the wrong plane. "Go to Armani, charge it to me," Jeremy says.

The next morning Jeremy's gone to work. Phoebe's at school. And the Girl's lawyer "is in court all day."

I take the dog and walk the half-mile from Jeremy's house to where the house I grew up in used to be. I am looking for more here than the girl Polanski raped.

The night my parents' house burned, I had awakened from a nightmare that my parents were enveloped in a sheet of flames. As I sat up, shaking, I realized the phone was ringing. My father was on the phone, sobbing, telling me the house had burned.

Now, as I walk, I can't remember why I didn't go right to him.

I cross Sunset, turn down Burlingame, and as I come near Marlboro I see the hill where we'd roll down the ivy. I don't want to go further, but I see the chimney, the thick rustic shingles of the high Tudor roof. I go left up Marlboro, hard hill to climb on my first bike, on up to the even steeper curve of the drive.

There are gates here now. And this is our house—it didn't burn down.

"That fire was in New York," my brother tells me.

He's here from Dallas on a shoot, and we meet for dinner in Santa Monica, not far from where we were taken to the merry-go-round so Tucky, our governess, could meet up with Mickey, the Marine who came over for naps.

Jeb looks like a giant cowboy: tall, stocky, bald, with sharp green eyes. "You know what was strange about that fire," he says, "was that the autographed pictures of presidents, the pictures that burned, were the ones that could be replaced. The ones of Kennedy, Roosevelt, and Truman were untouched."

The driver, Charlie, found me exactly where Judith said he would. As he opens the door to his 1983 dark blue Chevy wagon, he tells me he's a writer.

"So, tell me what you're working on," I say. He's driving me to see Howard Koch.

"I'm writing a screenplay about the Mayan sacrifices. They'd kill their own thirteen-year-old daughters, sacrifice them to the snake god. Quetzalcoatl."

Maybe that's what happened with the girl who got mixed up with Polanski. Did her mother sell Little Red Riding Hood to the wolf?

Charlie pulls a black curl through his silver earring, winds it out and round his forefinger. He is wearing an old purple velvet frock coat with his Mozart t-shirt, his jeans, and Timberland boots.

He tosses questions around when we're at STOP signs.

"Do you see big changes since you lived here?" He looks me up and down, eyebrows raised.

"I'm not sure." Is today's answer the same as yesterday's? "I think it's more like a big city, but I only really see what I remember," I say. "I block out what I don't want to see. I remember the roads the way they were when I was driving with my father, but he's gone."

"How long?"

"Almost twenty years."

"Does twenty years feel like a long time to you?"

"Not as long as it does to you."

"Do people know when they're old?"

"Maybe they know," I say, "when they're asked that question."

He's quiet. We've hit a long stretch of Sunset where you can still use some driving style and wind up ahead of the traffic. Not a tree, not a patch of green has changed. I can't remember the name of these dark bushes with the slender pale blue flowers. I sat sucking the honey out of them, watching, the day Billy Keane and Bobby Redford came over and played football on the new grass.

We're at a stoplight. He pops a handful of salted almonds into his mouth, tossing his head back.

"Thirteen," Charlie says, "is the end of sexual wonder, which is why sacrifice never really bothered anyone. Magic ends. Life gets practical. Why go on?" His voice is flat.

The Girl is thirty-something now. Can I ask her about sexual wonder? Time speeds. The question seems archaic today.

"You have this story here every week. Charlie Chaplin, Fatty Arbuckle, Errol Flynn—those are just the ones you know," says Howard Koch.

Howard used to race around in a wood-paneled convertible. Now he makes movies from a wood-paneled office that looks like a library. He knows more than a lot of guys who don't talk so readily. He's at ease with himself, which puts big directors and big stars at ease around him. "The great ones all have temperaments," he says. "You can't expect them to change."

"She really was just a kid. The mother was the real trouble maker. She made the judge angry, too. If it had been someone else, the studio would have got some top lawyers for him. The judge was thrown off the case, you know. That's a big story."

That evening I go to the opening of Judy Chicago's amazing "Dinner Party" at the L.A. Museum of Contemporary Art, the spectacular production that Harry Hopkins first brought to the San Francisco Museum of Modern Art in 1979. I gaze at this symbolic history of women spread out here in iconic glory, each place setting a different artist, and I'm pitched back to the Banner Party on the beach where we each brought banners we made. We tried to be the art. We became the assemblage, spending a dark, hallucinogenic night burrowing under one giant banner, slithering out at dawn.

The following day, I drive to La Jolla with Dennis Hopper to see a retrospective exhibit of his photographs, which have the originality of his crafty eye. Like old cowboy buddies, we eat at a western restaurant. It could have been old rustlers, old roundups we go over, careful, distant. He talks of his son, Henry. I talk about my grandson Justin's batting average.

We talk carefully of the sixties and of the art scene in L.A. thirty years ago. "The sixties started at the Pasadena Art Museum," he says, "with Walter Hopp's show with Hockney and Oldenberg in 1963."

The whole town was changing then; as one part of L.A. became more conscious and sophisticated as an artistic center, its core enterprise, the studios that put the town on the international map, was turned over to corporate investors.

"The big change in Hollywood started even earlier," Dennis reminds me, "it started with Jules Stein. He hired people like Wasserman who would run it like a business. The old guys knew how to deal with artists."

Did they really? Some of them understood the temperament, the nervy bravado one day, the steely need for isolation the next. Some of them sheltered all that, but also used their artists, seducing, coercing, possessing them, running them like racehorses. Today Barbra directs her own pictures, and Dennis can choose the parts he wants to play.

His hands are worn and rugged, but so are mine. "There aren't even any theaters left where you can show an art film, an independent picture. When I go to Europe, I'm blamed for being part of the American industry."

We all have our own way of seeing it. Dennis is idealized in Europe for being one of the first actors to bring the wild American cowboy into the twentieth century.

I see him more as the shy, pale young theater actor Brooke was walking with one evening (maybe it was in New York), just after he'd done *Giant* and hadn't invented the camouflage you need to survive fame. His eyes were wide and gentle. The two of them were holding hands and, I thought, defining love.

And then we talk about what I'm doing here. "One of the editors told me it's the story of the century," I tell Dennis.

"If this is the story of the century," Dennis says, "then the century has a problem."

"I'll get in touch with you Monday or Tuesday, possibly sooner, but I'd hate for you to be terribly disappointed if she's deciding not to tell her story. The only reason that she's willing to is she's in dire straits." I can hear the shrug over the phone.

"Is she fragile?" I ask. I imagine her still this little kid.

"She was tiny then," he says, "but she's a big woman. Very attractive."

"If she's uneasy, what about if I go to see her?" I could tell Silver I may not even remember her name. "If I meet her, she'll realize she'll be giving her story to someone she could trust. I could tell the story of her survival."

The next day Charlie is busy writing. Jeb's son, my nephew Zachary Schary, a guitarist with my father's eyes and my brother's cowboy way, speeds me out to the Valley in his maroon onyx '66 Mustang convertible. The high school the girl would have gone to is gone. The boy on the phone who works in the junior high store says he's not sure where the high school is, or how late the shop is open. He just works there, you know. He says they might have some old yearbooks from '77, but he does-n't think so.

Larry Silver calls to say, "My client might be interested in The Big

Idea. You can't tell anyone you've gotten this far. You have to give me your oath and I have to trust you."

The oath I give is not to use her name or tell where she lives.

"So I'm going to patch you through now." Thirty seconds of silence goes like a month until he says, "Jill, this is —— ——."

"Hi!" We laugh.

"Hi." She does sound like a kid. "I've never known what to say about it." She has the light just-south-of-L.A. drawl, not quite a twang. "What if it becomes a movie? I don't know how I feel about that."

"What I think I want," Larry adds, "is to be in on all your conversations." He feels like a man who wants to "watch."

"Sure," I say, "if that's okay with ——. I find it harder talking in front of a guy, but that's up to her." I may learn something from her.

"I don't mind," she says. "So what do we do now?" she asks. "I just don't want to do anything that will be hard for my kids."

She asks what I'm really writing about. I say I'm writing about how some of us reinvented our lives in the sixties. I've written about being raped and dealing with mixed feelings.

Then when I point out some of the other people I'm talking to about the sixties, she's concerned. She says, "I thought this story was all about me."

I've been told by another detective, a guy in Hollywood named Warren, that I'd be smart to drop the Polanski story, stay out of it.

"I'm not that smart."

"She's living in a place out on the fringe of Santa Monica around Federal," the detective says, "but I'm not sure what name she's using."

"That's where nurses at the old veteran's hospital shared houses during World War II," I tell Charlie as we drive in to Hollywood to meet Warren.

The word you use to describe where you're going is very important here. You go "in" to Hollywood; "over to" means the Valley; "up" is one of the canyons; "down" is Sunset to the Coast Highway; and "out" is to Malibu.

"Maybe you took speed to bring back that war panic." Charlie likes to think of me doing drugs. It shows a hip fallibility. Or maybe it shows you can do drugs longer than he did and live forever.

Warren says this actress named Lisa is absolutely the right girl. "It got me a lot of good parts," she says when I call her at her apartment off Mulholland (on the Valley side). "Actually two," she said, "they weren't big ones. But I didn't discourage the idea. To be linked up any way at all to a famous name in this town is useful. You can always do something with it. I haven't figured out what."

Not yet.

Lisa mainly needed to talk, which I understand. She had been alone in her house for a week. She wasn't certain if her husband was in jail or had forgotten where they lived. He had a two-hour memory range that he hadn't found a new chemical to enhance; and if he did, he'd forget where he'd written down the source's phone number. And it would be unlisted. She reminded me that all good L.A. phone numbers are unlisted.

"Do you know any good agents?" she asked me after we talked the third time. "Maybe we could work it into a screenplay together."

"I don't think Lisa is the girl," I tell Charlie, "she doesn't have that permanent edge in her voice."

"What do you mean, permanent edge?"

He'd have to listen to the girl to know what I mean.

All weekend a Santa Ana rages round the city. The coyotes sound closer, wilder, as more of the chaparral, their homeland, burns. Down in the shoplands the air's smoky and you can see the burning patches on the mountains.

I meet Army Archerd and his wife, Selma, at Drei's on Bedford. Just opened that week. The owner, Victor, tall and French, spins down the aisles ecstatic over the rush. Drei was a friend of Polanski.

Army saw Roman six months ago. "Roman feels abandoned. Jack's dropped him. Bob Evans. All of them. Sure Roman's arrogant. That has nothing to do with his work. So he's tough to work with," Army shrugs. "Some of the best talents in the town are tough to work with."

None tougher than the lawyers.

On Wednesday, I'm up early making a birthday cake for Jeremy with Phoebe. Lots of fresh coconut and lemon juice fixes the frosting flavor. The red crystals, the rose in the center, the almond toffee, the raspberries and the strawberries. "It's brilliant," she says.

When we come back from a walk with George, the golden retriever, there's a message from Silver. I call back at 11:55. His office says he's on a conference call, "he'll be a minute. Do you want to wait?" "I'll wait." He's still on the other line.

We do the jockeying, the game, the "what are we going to do here?"

"Why do I think I trust you?" he says.

I don't tell Larry I've heard the thing to see is the testimony.

"So, I have a friend who says you're a great lawyer."

"Do I know him?"

"He's a lawyer. He's heard of you."

How long will we play?

He patches in the Girl. This time I talk to her for maybe ninety seconds. She has a hesitant, shy kid's voice. She speaks L.A., which clarifies only when you get out of town long enough to miss it. I talk to her long enough to believe she's real.

"So," she asks, "what are you going to do?"

"I thought I'd come to see you."

"Maybe you could send me something of yours to read. And we could talk about questions so I think about them. I'm doing this so he can come back and just let it all be finished."

I sent her a couple of pieces I'd done for the *New York Times* and a few ideas I felt we could talk about.

Then when I call on Monday Larry can't talk to me. On Tuesday, "It's a week from hell."

"You've had that week. Listen, I can't just sit around out here." It's not going to happen.

"Come in tomorrow," Larry says, "I'll have things for you to read no one's ever seen."

I am waiting at Nate 'n' Al's. Delicatessen is one of the things Polanski has told me he misses about L.A. I taste the rye bread, the new pickle. I tear into a bagel.

I am waiting for my father. He's got on his cardigan and the hat he wears coming on Sundays to pick up everything for our deli dinner. My father must be having trouble parking the wagon. He always brings the wagon on Sundays . . . I freeze.

Images skid by. Jeremy's red room, cactus garden, Josie's car. Sit very still. The Wilshire Palms, Cary Grant on the balcony, tanning, reading a script. Pale blue cover. Arrowhead.

No wonder they're called "mals." This one is not petit.

Slowly, too slowly to feel safe, it's going out of sharp focus, fading like a shuddering, uneasy come, and I'm unclear. Time and place are off.

I look for my father. Tears. Don't. Look for his tennis sweater. Lennie's friend's shop. Go there, they'll call. Smarter move—go nowhere. Yes!

I see Lew Wasserman. His hair is platinum today. He always looks like a classy piano player at a good club.

Wasserman is my father's agent. Even when he makes house calls, he wears a black suit, white shirt and a black tie. He's like a tough editor, the art dealer, the teacher. Every artist needs one around. He and Edie stay pale so you'll never take them for westerners. He smiles at me, kisses me, say he hears great things about my son. The confusion may be his, but Wasserman has never been confused. "Wasserman was the first one to get an actor a piece of the action. He got James Stewart ten percent," Alan Ladd Jr. told me. "That was probably the end of the studio system." A subject that alternates in consideration with Sharon Tate's murder as the end of the sixties.

This information comes on like a slide even as I am lost here in midimal time.

I am shivering, numb, but I know to look in my Week-At-A-Glance. Yes, I am to see a lawyer named Larry Silver at 3:00. This may be about our divorce. I thought Bert Fields would do it, but he was at our wedding.

Finding my way is no problem. I'm blank right now, but I can still tell you every street in Beverly Hills.

There are different memory road plans. For example, there's the grid system, where characters interlock. You have to make clear choices here, and now here again.

I'll take the axis, a direct route, but I'll have enough shapes and distraction to make me think I'm on a wandering path. Wandering paths are a favorite in England and Japan; you can feel you're lost or go directly. A good museum is planned like a wandering path so you can feel you're strolling along with idyllic stopping places. You know where you are, the way you do on a fine axis street like 57th, but you can also be lost at any moment exploring Regent's Park or Fifth Avenue or Sunset Boulevard.

On the way down Beverly Drive, I pass a Hasidic rabbi in pedal pushers. Cannot be a bad sign.

I go into the lawyer's office. The walls are gray, carpet a sea gray. No, it's not the divorce. It's something to do with Slipkovitz. Yes, that's it. Am I doing a story for *NYPD Blue?*

Don't, above all, appear confused. Do not ask revealing questions. Try to gather clues.

Silver may dye his hair. Under his glasses one eye wanders. Maybe that's what makes me feel he's disinterested, not paying attention. There are piles of cardboard boxes and cartons, one turquoise office chair, five brown ones. Like the new cars, the chairs are all shaped like eggs. It's sort of high-tech in midlife crisis here. There are crumbs and ashes on the glass-topped wood table and a small cobalt blue Indian wall hanging.

It's not an MGM set. Doug Shearer did not do this.

I leave the office. I have taken notes from pieces of testimony Slipkovitz had me read in front of him.

"Listen," I'm still smart, just addled, "if I'm taking notes, why can't I just have a copy? It's only one page."

"Because if you use it, I can always say you made it up."

A chance here for a clue. "Use it for what?"

"Don't play games with me. We both know what's going on."

"Sure," I say, tough. I'm copying the dialogue here.

Silver says he'll call me tonight. I tell him I've got other fish frying. That's always safe in any scene.

Feeling daunted, I'm sitting on a stone bench outside this skyscraper on what I know is Wilshire. Worse than being seen walking is being seen not driving. It does something to the tone of your walk in Beverly Hills when there aren't keys to a good car in your pocket. Not to drive in L.A. is to lose virility. I consider hitching. Would anyone know what that is?

A young man comes to the curb. I know the car. "I think I'm doing a TV series, yes? And you are?"

"I think you're tired," he looks at me, puzzled. He has a Mexican accent. "I'm taking you home."

Phoebe cartwheels down the hall to meet me when Charles drops me off at UTA, Jeremy's agency. Phoebe and I work on some drawings in Jeremy's office as he twists the phone cord, taking calls, checking papers slipped in front of him. He's exactly like I'd like to be.

"I haven't been here in a while," I say.

I can't tell him how it feels to be sitting here, watching him so cool and alluring. He is distinctive in his distant style, achieving, then moving beyond the most elite Hollywood manner. That's how I see it.

Not that, like any self-absorbed mom, I drive him crazy, and that the way to deal with me is distance. After all, isn't that how I have dealt with him?

Later, there's a message from Silver. As I wait for him to answer, I wonder if I'll make this first draft of the story before I short-circuit. I thought
the needy panic, the hysteria I saw in my mother when she was getting
work done before an exhibit, or in some stars before a movie would wrap,
was a creative prerogative, maybe the only way to set yourself up for the
last burst of energy. But I can't dare that. Chill out.

"I don't think we want to do it," Silver says, "we've got a TV offer
now." Pause. "If you agree to publish after the TV show, maybe we'll see
what we can do."

"Larry, don't play games with me." This is aloof. "I'm going ahead."

With what, I ask myself.

What's startling is when someone not only remembers you, but comes up with a quick image that gives you a new take on who you thought you were. Even as a kid, Bob Redford drew new compositions and took unusual angles on each group of figures he'd draw.

Bob is sitting in the back of a restaurant I've never heard of, at a table you'd never see, in an angle of glass high in the trees looking out over a stretch of the city I never knew. "I do know you very well!" I say, surprised.

And he's always taller than I remember. He was a tough, kind of edgy kid, an outsider like me when I first knew him at Brentwood Grade School.

I'd been sent to public school for the first time. I was here because my father took stands. "That's your American responsibility," he said.

I thought we had to leave our private school because at the parents' meeting they decided to expel the kids of blacklisted writers. We couldn't go to other private schools in L.A., because they either had quotas of no more than 20 percent Jews or show business students at any time, or none at all.

This is how my memory works today. I write this down. I know it is wrong after I see it on the page. This is only part of the story. We left that private school because the head of the parents' group thought Robert Mitchum's sons ought to be expelled because Mitchum had been busted for smoking pot.

Ed Lasker, a lawyer and one of my father's best friends, was going to the meeting. My father said Eddie was too hot-headed, so he'd go. He could keep his temper. My father "slugged the SOB" and we were, of course, in public school the next day.

This is how my father told the story. He liked to say "slug" and "SOB" Looking at it even another time, I figure he took a firm stand, then, once defeated, stalked out.

Since there had been letters threatening kidnapping, we couldn't just go to school in a station wagon driven by our governess, who could at least pass as a mom. We had to go in the limousine. Other Hollywood kids have talked about having the limo drop them off a block or two away. Donovan, our driver, wouldn't do that. He'd just gotten out of the Army and needed the job.

The spaces on the playground were huge and circled with high wire fences. At school they tried to teach you to be good at everything. If I'd been running schools, I would have let you do what you were really good at and left it at that. No phys ed. No math for me.

I'd sit on this bench, picking off flakes of gray paint. They couldn't repaint because of rationing. I looked up at the wire fence. I couldn't imagine how you'd get out. There were no trees, sparse palms around the outside. I noticed Bob first when he was playing ball. He'd gone after a ball no one else could have caught.

Even in the middle of team sports, he was detached, the way I was. He'd rub his bristling reddish hair and size them up.

I envied the childhood I thought Bob would have. It would be like the childhoods you saw in textbooks. Most arithmetic problems included pies. We'd remember numbers given familiar images. We did not have homemade pies. Perhaps if the books had pictures of kreplach.

"If Jane's Mom bakes six apple pies, and Dad lets Dick mow the lawn seven times, and Uncle Fred says they each take twenty minutes to finish, who is ready to take Jane and Dick to the beach first, Mom or Dad?" I couldn't figure out the answer. I was picturing the simple intimacy of such a life.

Inside the classroom, Bob sat away from others and looked out the window a lot. He was fixing his attention out of here and into his head, which I understood.

I watched him without talking and managed to sit near him without catching the attention of teasing kids or teachers, which was a trick and entirely due to his subtlety with moves and placement. He could be in the center of the room here answering a question, disappearing the next. "One day I just couldn't bear it, so I went to the boys' room, opened the window and left. They put a monitor on me after that."

Here with Bob tonight, he is an artist. He listens with the intensity of the artists I've known—with my mother's early sharp perception. And like men who are artists, once you are alone with them, they can match you even in talk.

"I was telling Stuart how I'd watch you draw action—and how still my drawings looked next to yours. He said it all comes down to the masculine and the feminine."

The arrow and the womb, waiting, is what Stuart actually said. And I am remembering, as I write this now, a story by John Updike. And I've found it, here. "These pairings," Updike said, "are like the Royal Albert Hall, round, capacious and rosy, and the phallic spike of the Albert Memorial."

I could place them, I was pleased to see, as I read. They are on the other side of Hyde Park . . . how long it took me to remember Hyde Park in London is not F.D.R.'s house.

When the teacher was occupied in another direction, without talking about it, Bob and I drew. It went like this. I'd draw one picture, usually knights or early Americans in an important scene. You had to put men in drawings, or at least buildings, to convince boys you were good at art. I never did figure out a way to be interested in drawing men sitting in plain business suits. Bobby Redford's pictures moved. Legs were really running. You could see the arms reaching back. My pictures were pretty, but they stood still.

At private school, Sue Sally Jones and Jane Fonda had had the patent on horses. No one else could draw horses or try. "I'll draw dresses," I'd agreed when we were laying out the territories. "Fine," Sue Sally had said. This was like someone saying "I'll take Catalina Island." Who cared about dresses?

To know an actor, look at his hands. He's learned to stage his face. Bob's hads are young and wiry and worn all at once. Tense, lithe and bow-legged fingers from riding a range of pencils, brushes, gestures. I'd guess he gets bored faster than the characters he's drawn up there for us, but then, he talks like an artist. You can see him roaming around in his head over the lay of a sentence or the shape of a scene, the speed with which he'll want to get to a story's point. He listens and watches you like an artist. In which way will you apply to a character? As John Lahr says, we all scavenge each other; we are each other's best material.

"You live with the dark and the light, so you use the dark. Put it in the work. All my best work is personal, like *The Milagro Beanfield War.* The great villain now is real estate."

"Yes," I say quickly. I can talk about the land wars that are going on now in Malibu, but I don't remember this movie of Bob's, and how can I say that so I told him about the memory thing, windup with this bit.

"Sometimes I'll say I want to see a movie, and Stuart will tell me we saw it last night. I won't know."

"Then you may not remember," Bob says, "that the first time you invited me over to play, and we tore up the grass? Well, when I was sched-uled for a tryout for a play on Broadway—I was terrified your father would remember me and I'd never get the part."

"But that's not what my father remembered about you," I say, "he remembered you were shy and had talent." If you know someone as well as I knew my father, you can play a safe hunch on what he'd remember. Memory goes more on hunches than you think.

"He was a nice man. He understood actors."

"I think he understood actors because he understood artists," I say. "I couldn't bear how much of himself he gave to my mother, and now I watch my own husband do that with me."

We speak of our children and political influences. "When I was thir-teen, Richard Nixon gave me an award," he says. "I remember feeling he was cold and wooden."

Then we talk of kids we both remember from then, of the first girl he kissed when she was ten, of the juvenile delinquents, "kids who could get you into bad scenes."

"They looked like John Derek," I say, "so you didn't care."

"They were fine until they'd go near your kid sister. You could deal with them. Then, too, most juvenile delinquents were more interested in your mom than your kid sister. The idea of kids and sex was gross," he says.

And we move on to the sixties, because that's when the story I'm here to write really began.

"It wasn't so much when the sixties began as when the fifties fell apart. But the Beats weren't embraced like the hippies were," Bob says. "I think it really goes back to the quiz shows. Remember how we felt? How Van Doren shook our belief system when he crashed and burned? It's hard to imagine an America where a guy who cheated on a quiz program shook us as badly as a violent act, but this was moral violence. Decibels so low we'd shrug it off today."

Bob didn't want to be kept in, but he had ranged far enough in those early years to be able to work with a disciplined vision later. "The big changeover has been in the feel of fame and creativity. The sense of being an artist is different from fame. Responsible fame requires playing a political role, but you still have to have a broad enough base of privacy to ramble around in."

"Polanski had never had it. He was like a kid in a big candy store."

"It was sad to watch people get sucked in. When Manson hit, some bubble burst. But Kennedy—that was the shock. People walked around numb."

We order finally. I'm not paying attention to the food. I think Bob keeps his privacy by talking mostly about you, but when he tells stories in scenes; setting them up, then standing back like a director. He talks of his friendship with Carol Rosson, whose dad was blacklisted. How he put himself through hell earlier in Utah. "When I was sixteen, I was going out with a woman who was twenty. I thought I'd be a lawyer, went into pre-law.

"But my sixties began in the late fifties with a trip that expressed the whole time—wide open and raw. I took off in a car with a woman who understood I was freewheeling—an artist. I came back, alienation a shell around me. I loved the art, but couldn't connect with the movement. I loved—almost survived—the spirit.

"The sixties came to be about support; if we band together there will be power. What I was doing was about the opposite. I was giving up that freedom to be with someone. I also loved what I was feeling. I didn't like classrooms, the group behavior of sports."

"I'm just surprised," I say, "that you remembered me."

"I'd see you come to school in the limousine like a princess and stand there so aloof and unapproachable. I'd want to talk to you because we both liked to draw."

I never saw myself like that. It was so excruciating to be seen being dropped off, and watched. I knew I was despised for this distinction, so I didn't try to show off or push my way in. "You were observing," I say. I have forgotten this cringing, despised pale kid I was.

"Aloof has some strength," he says.

"I like that," I tell Bob.

This is going to be my new style. Unapproachable. How much work you can do when you are unapproachable, and how useful it will be to be aloof with Silver and the Girl.

34

Stephen Kabak is who you call when you need to know something on record. He has the lead to the right city offices. He would be played by Ben Kingsley.

I am going down to the Hill Street courthouse to get the transcripts of the testimony no one ever saw. Gittes went there to get the papers in *Chinatown*. I'm standing outside the car down here at the library when suddenly I am the person I was then, the person I was in the sixties and seventies. When I was a kid, I wanted to be Helen Gahagan Douglas. I wanted to be serious, to be political. That was another kind of importance, not like being a movie star.

I scramble with Charlie down three basement levels into the archives of the courthouse. Movie posters are taped on walls: *Disclosure, Night of the Running Man, Imaginary Crimes*.

"Did you see the man who works here?"

"It's the Sequestered Department," we're told.

"I know, that's why I'm here. Is Cameron the guy I'm looking for?"

The man who doesn't work here asks another man walking by, *"La señora quiere ver viejos archivos del tribunal. ¿Necessita hablar con Cameron?"* Even though I can't speak Spanish, I get the drift: This woman wants to see the archives. Does she have to talk to Cameron?

"No, necessita hablar con Walsh."

No, she'll have to talk to Walsh.

Walsh is the name of Gittes's guy in *Chinatown*. This is perfect.

"So, can I see him?"

To Charlie he says, *"No, no puede."*

"No, he's not here."

Charlie tells me, "He's just stepped out."

"¿Puede ver el archivo?"

"So when is he likely to be around?"

"Es muy raro." I sit down, discouraged. *"No sé si vayamos a encontrarlo."*

"Hardly ever."

Charlie talks to someone else. *"Ella es buena onda."* "She's a kind of good guy," Charlie says (or something like that). They look back at me. The court guy leaves for about fifteen minutes and returns with the document, which resembles a screenplay. He hands me the testimony. Even when it's a terrible story, there is this excitement about seeing something no one else has seen. This is probably called prurient curiosity. I start to read it, to take notes.

At approximately 2:30 on Thursday, March 24, 1977, she is called as a witness before the Los Angeles County Grand Jury. She raises her right hand and solemnly swears that the evidence she'll give in this matter now pending before the grand jury of the County of Los Angeles will be the truth, the whole truth, and nothing but the truth. So help her God.

She is thirteen. The man asks all these questions in a cool, matter-of-fact tone. When asked if she lives with her mother and her twenty-year-old sister in Woodland Hills, she says, "Yeah."

She's advised that she should answer "yes" or "no."

So she says "yes," and she tells her story:

This is what I want. "Charlie, I have to have it."

He talks to the guy who looks me over, shrugs, *"No puede llevarlo."*

More talk. Then there's a conference with someone else. *"Díle que regrese manana. Quizás tendremos alguna noticia."*

Charlie can come back. They'll have a copy tomorrow.

I stay up tonight reading the court transcript. I imagined myself at thirteen saying some of this out loud to a group of people. They are mostly strangers, which is at once better and worse.

When your first sexual experience is bad, you can be destroyed by it, because every time the sensations of sex come up you have that image. If you're lucky and work it out, it fades away. Secrecy maintains the sexual force of the iconic images. I would guess there's this girl out there who sees Polanski for an instant every time she gets off. The worst thing with this is you feel bad about yourself, certain everyone else sees the face of her lover in her arms.

A legal advisor I talked to said, "At thirteen a girl is not emotionally ready to handle this. She won't be just fine."

Maybe she's fine enough to know it won't be right for her to risk more by what we call this sharing. Maybe she's a fine enough person not to do a public tell off of her own mother. Maybe she's one of the sixties children who has grown up. Will I have the grace to let her be?

I call Silver back. I am very clear now who he is. "Another thing we have in common besides rape is asthma." I need my inhaler.

"But my client doesn't have asthma, don't you get it? She said she was having an asthma attack so he'd stop. Polanski didn't pick up on it, because she thought it would frighten him that she was having an attack. One of the things about rape is that you try to resist. Let me put it this way," he slips into the legal voice, "in order to demonstrate that it was rape, you have to prove that you tried to do something to stop it, and because she did pretend that she had asthma, that was proof enough that it was rape."

How did she know to do that? Maybe it's built in. A girl I know told her stepfather she had AIDS. "My mom's nephew died of it," she told him, "so he'd believe it was in the family."

"So where are we?" I ask Silver.

"You're a nice lady. I don't want to disappoint you."

"Then don't. How often have you seen Polanski?"

"Only a couple of times. I hate his attitude about women." He stops. "Look, I'll talk to my client."

I'm sitting here on Jeremy's patio looking out at Phoebe directing the golden retriever, and I'm listening to the girl's voice. It's Saturday. The girl is easy, open—that's the surfer cover. "I'm just worried about my children, their privacy. And do I have to go over what happened then?" We go over all the fears and restrictions. "What if it becomes a movie?"

"We'll think about that when it happens," I tell her.

Larry calls. "She wants to do the interview." Then he has to take another call. I hold and hold and hold. "The music is Aaron Copeland," Larry says, "could be worse. She'll do it, but I want to protect her rights."

Three days later it looks like arrangements have been made to meet the girl. The day before I take off to meet her where she lives, Larry calls again. "This happens to be a coincidence," he says, "she's coming to LA."

I talk to the girl again. "Larry says you want to do the interview. What a great coincidence that you're coming here."

"I really don't want my family mentioned or my name or where I live."

"Of course not," I say quickly.

On Tuesday, Larry calls again. The girl wants to change the place to the Disneyland Hotel. She's bringing the children there. They'll go off with their dad. The first priority is the kids. Her husband probably hates the idea.

I call the Disneyland Hotel and talk to a guy named Joe for directions.

The next day Silver's changed his mind. There are new demands. This time I say, "We'll have to pass."

"All my clients promise me I'm going to be famous," he says, "and nothing ever happens."

It happened. I wrote the story, and George Plimpton included it in his anthology *The Best American Movie Writing 1998*. I never used the Girl's name or said where she lived. If you ask me, I say I've forgotten.

But a month after *Vanity Fair* published my article, she came out and told the world. You can't get this close to fame and not bite.

35

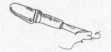

We are going to New York. Then I am going on to Connecticut to see Johanna and Stuart is going to Boston to see his daughter Susan and her mother, Margaret, Stuart's first wife. Margaret and Stuart have lost one of their grandsons, Kenneth, who was found at the bottom of a cliff in Montana. Stuart Jr., Kenneth's father who is an artist, an original, and Kenneth's brother, Nathan, will also be in Boston.

Stuart and I are fighting. "Maybe we always fight before we're going to be apart? Is that true?" I say.

"I didn't notice we were fighting," he says. He's in his tweeds, back from the Regent's Park walk.

"Maybe we do need the space," I say.

"I don't," he says, "you're just working up to a fight because you have just about finished your book."

Each chapter is laid out on the dining room table. Angie, Geraldine and I met last Sunday and they helped me lay it out. Angie read the new beginning of her book, the first real work she's done since her daughter Delilah was born. Geraldine's *Tempo* is ready to send out. Svetlana is posing for artists again. "I hate to write," she said. "That," Geraldine says, "doesn't stop me. I hate to write."

The night before we go, Laurie comes up for supper. "News!" she says.

She has been asked to go to Russia by a man who is the benefactor of an entire village, the last place where they do those paintings on black enamel. "They asked me how I make my pencils," Laurie tells me, "so I thought I ought to ask how they make their brushes. 'First,' I was told, 'we catch a squirrel.'"

Again Stuart and I are looking over chapters.

"What matters is I can remember exactly where each scene is."

I was triumphant, until it occurred to me to change several scenes around. Which is how it would go if I switched images in Steven Rose's memory test.

Arguing generates adrenaline. I also am missing my children. "They're hardly in the book," I tell Stuart. "There's just this one last bit."

I call Johanna. "I really haven't been there. I'm reading over the book and I haven't written about your life—your kids, your husband."

"Maybe you haven't written about it because it's my life," she says. "So, are you at the fighting-with-him stage?"

"How do you know?"

"Put Stuart on the phone and I'll remind him you must be finished if you're being awful and sending out for pizza."

"I just made brisket and applesauce."

"So it's another week," she says.

"Your children might hate being in the book much more than they mind being left out," Stuart says. "They're more private people."

"Everyone's more private." I turn on him. "What about you? Do you remember how I just rip pieces out of your own writing when I can't think of how you'd say something?"

"Yes, but you don't want to remember right now that I gave my journals to you."

"But why? And what about that, how you've just tossed your whole life down the pit of my neediness with this illness, with my writing, always my writing. Do I write about that?"

"It isn't my whole life. And whatever you talk about, you will write about. I knew that the minute I fell in love with you. This is our reality. You'd never have stayed with a man who put money and power first."

"Do you have to put it first to have it?"

"You know the answer to that better than anyone," he says.

"Do you question that choice?"

"Jill, why talk about it? Could we go back and change that? Would we?"

I'm digging deeper into an area I don't know how to write about. "Think of the women, the writers I knew in New York in the seventies." I can only see, now, looking back with my adaptive memory, how close we were. "I can't think of any of us who write about our adult children."

The look I get now over the plate of brisket is, "Would you know? Do you talk? Look out for each other?"

"Don't," I tell myself, "think you'll be going back to that New York, any more than the L.A. you saw while researching the Polanski story was any of the L.A.'s you'd known."

But I wonder where my children are in this consideration of these last few years. It's not because I don't care, not because I've forgotten. It is because this is the last taboo. Am I stepping across that line? We go after our parents, our mates, particularly former ones, but what of our children? What of this deep, complex time in this most perplexing relationship?

Do they all hate what we do, and that we're doing it for more years than ever before? Did they tolerate the writing as a single mom's job, something that would sometimes pay the rent?

I understand my own mother only now. I may have my talent, but I caught the obsessive drive from my mother.

Before I leave London, Phoebe calls me from L.A. Like her father, she's fast on the phone. I communicate with her best by making small books. "Grandma, there's a Britannia Bear Beanie Baby," she tells me, "I really need to have it."

The Brittania Beanie Baby is symbolic of everything I need to be. Scratch the book.

Fenwick's doesn't have it. Selfridge's has no idea where to find it. We call Harrod's, who gives up a number to call back. "You can try Hamley's," Corinne suggests. Hamley's has a waiting list. We get the Harrod's number when it isn't busy. A message says, "If you're American, don't call because we're not selling to American distributors." Or grandmothers with American accents. I fly off without the Beanie Baby.

When I call Johanna to say I have arrived in New York and that I am staying at Bruce and Lueza's, she reminds me Jeremy might be coming to New York. He might be bringing Phoebe. Or might not. And we might all be going trick-or-treating together. Or might not. "And," she says, "he might drive you to Connecticut."

"Jeremy drive me?"

"Yes, Mom," she says, "he drives."

They've been talking. I like that. Look how long it took for me to really talk to my brother, Johanna reminds me.

There might also be a confrontation, or might not. The images of their past must be so powerful. How else can they get over the anger?

The night before Stuart goes to Boston, we go with Ted and Vada to an event. I see faces I know and can't place. Do they know I don't mean to not remember them? And how do you go up to someone and say, "I'm crazy about you, who are you?"

Then suddenly, here's Brooke. Her face is glad to see me. During these last years, I thought we'd met sometime in the sixties. Is it the music playing—and I don't remember what it was—which creates the connection; or, as Schacter might say, electrifies the engram. But I suddenly see Brooke and myself and we are six years old. And I know the image I saw in London of our children lying on the floor in my house in Santa Monica, has stayed with me because it reminded me of all the times Brooke and her brother and sister, and me with mine, played in the barn Brooke's parents built for the children next to their Brentwood house. After we'd worn ourselves out, we'd flop down on the floor together, drawing and making up stories.

"I've known you forever!" I say and we hold each other tight.

The next morning the phone rings at the Gelbs' at six in the morning. "Who's this," I say into the phone.

"It's Jeremy."

"Really—you?" I want him to call so much that by the time he does, I've worked so hard to drop it, I can't place the voice. It's like when my father would call when he'd gone to New York. I'd always dash to pick up the phone, to be there first, to catch it before my mother, and surely before my brother or sister. It's the way it is when, at last, you meet the hero, and stand there dumbstruck.

There are no games here; he presents the agenda. I've won the lottery. We'll have dinner tomorrow, take Phoebe to FAO Schwartz Friday morning and to the theater Friday night. "I'll drive you to Connecticut Saturday and we'll go trick-or-treating, and Phoebe might stay over."

Actually, this is a game, and the game is minimalism. Can I keep the conversation more direct, more simple than you can? "I can't do theater. Sorry. But I'll see you tonight." I stop. Is this too much? The main thing is not to be boring. They talk faster. I'd shape words into sentences before I'd present them to Anatole or Leo Lerman; sentences that might lure them when I wanted to catch their attention. Don't fool yourself. They were just as fast. Anatole picked up the phone and said, "Yes." Leo would say, "Four-thirty. That's fine."

By the time you said "Fine" back, he was off the phone. "Phone" is not the same as conversation. "Phone" is not to be confused with talk. "Phone" is like Morse code. We used to talk about people who gave great phone. You'd talk and never have to see them. Giving phone is not the same as phone. You don't give it, talk it, or do it. "Hello" is over. "Goodbye" is gone.

With e-mail, you don't even have to talk. The typewriter meant you no longer had to show your handwriting, and now e-mail means you don't have to reveal the style of your letter paper. Tiffany's on Bond Street no longer sells stationery. Cicero would be so relieved. Writing's over.

I had wanted to be at Jeremy's hotel, to meet him there, but Stuart would have told me that might seem overeager. I should wait until Jeremy calls. You can be the perfect mate, but the exchange between parent and child is different. The complexity passes down through the genes. Maybe I'll just go over there.

I throw off my jeans skirt. It's New York. Black will be the thing.

"Does this look too odd?" I ask Lueza, taking off the wired black necklace with the spacemen. "Trying too hard?"

Jeremy called the day Phoebe was born and talked about how everyone was driving him crazy. And Stuart said to me, "Isn't it great you're not there?" But I'm different. Maybe too different, so it's best I wasn't there, coming up with distinctive ways to be helpful.

This is exactly like waiting for a date to call. No, it's not. It's more, and I remember what it's like. It's like waiting for your parents to come home from a trip. Or wondering if they'll come to see you read a story at school. At night I'd wait to see what my father thought of a poem I'd leave on his pillow. Would there be one of his notes on the yellow-lined paper folded on the breakfast table in the morning?

How many times in the self-absorbed years of the sixties and seventies did I miss the cue? Maybe I won't miss the cues with the grandchildren.

I can remember loving my father because he understood how writing went, for knowing the movies I would love and talking them over with me on the same level. He gave me Ralph Lauren pajamas the year he knew he was dying—a gift he couldn't afford. Did I let him know how much that meant? Did I know how much it meant to him to give me something I'd like, to know he was still in touch?

Do I know what Jeremy would like? Or Johanna? Is that another reason I keep close to these writers and to Laurie, so I can grasp their generation? This isn't so much memory as keeping the spirit supple.

The phone rings. "He's here!" I grab it.

"Where are you?" Phoebe says.

"I'm here."

"You were going to be here by now." Grandchildren are the reward for all the cool years. You cannot be too much there. You also cannot say I forgot.

"I will be right there." I pick up my black scarf and throw my coat over my arm, and I'm off.

No lights have been so long, no traffic so heavy. I hear the Englishman's voice. "You could walk this faster."

You can't walk that fast in New York because you'll see people you know; between Sixty-Seventh and Fifty-third on Madison, I'll hurt twelve people's feelings. "I didn't know you were in town." "I'm not, actually, but I'll call when I am." I dash by, "I can't ever remember to call."

"It's so smart of you to wear black," Phoebe says, "that's what I always wear in New York."

Phoebe is seven, an early preteen. You're not just a child now, you're categorized by your economic power. A mid-preteen doesn't have enough allowance to get an older brother or sister to pick up a video she isn't supposed to watch. In a top executive partnership where both parents work, the post-preteen films her own video.

I can't take my eyes off Phoebe, who is sprawled across a sofa in the living room of their suite. Her black fur coat and hat, which have the look of a turn-of-the-last-century sealskin, are draped around a purple Teletubby. Jeremy is on the phone. Blair, Jeremy's fiancée, is warm, friendly, and easy, with short dark hair and the organizational momentum of someone who does well in L.A. She can get a few people who have very different agendas to appear to pull it together for a while, to move in the same direction, without inciting rebellion. Jeremy has the crisp, tough attitude that gets Phoebe to consider tossing her coat on the floor and saying "no," and makes me want to have a talk about where we're going and what we're doing later.

Blair sizes up the characters, gets our coats on, gets us out the door and down the elevator in minutes.

What I do best with Jeremy is attend to Phoebe. He has a twenty-second-century mind. This is a Hollywood technique he's mastered: from the beginning of time in Hollywood, the idea is to be ahead of you. New Yorkers, you'd guess, would be better at it, but they're not.

The day in New York is bright blue. The streets are filled with women I understand.

"I'll meet you at Schwartz at a quarter to nine," Jeremy says.

Do I tell him I don't think it opens until ten? How do I remember that? I tell him.

"How do you remember that?" he says, then adds, "I've got a buyer meeting us there early so I can shop for Johanna's kids."

Phoebe will have FAO Schwartz to herself for an hour. I remember Miss Frenault meeting us at Saks and having three dressing rooms filled with clothes to try on. It wasn't a big deal.

Phoebe's first question is, "Do you have the Furbies?"

"We're sold out, but they should be in tomorrow."

Phoebe has been given a budget for the day. She keeps to it, knows she can't have everything. I never knew that. And then I told my kids they could have everything, and I never made rent.

Jeremy and Blair discuss which latest computer space radar game Justin and Ethan would like. Phoebe wanders about, trails the coat, dances. "I should have my hair done," she says after a look in a mirror. "I saw a place on the way to dinner. We can go after we're through here. It was cool."

"Do you know where the place was?"

"No, but I'll know when I see it. We'll just taxi up Madison." She's dropped "Avenue." She moves from L.A. to New York with perfect cool.

For Halloween, she's wearing a Ginger Spice English flag chemise dress Jeremy bought at auction. She takes a magnanimous preteen view of Alice's pink butterfly outfit. "She might like a wand," Phoebe says. She wants it clear that she's not into that kind of thing.

"Do you think I should wear a wig?" Phoebe asks as we start up Madison. I want to show her the new Chinese shop, Shanghai Tang. "My mom got a red silk shirt from there," she says. "I don't think they have a kids' department."

It's important for me to be right with Phoebe. If she likes me, it will get around to Jeremy that I'm not brain-dead. You want your kid to think you're smart and, most importantly, you want him to trust you with his kid.

"I'm sure they do," I say. Michael Chow takes us downstairs, where Phoebe picks out a red satin jacket.

Life is a circle. I see myself smiling in the red satin Chinese jacket my children's father brought home from Hong Kong when he was a naval officer.

"Perfect." Phoebe regards herself with a calm, distant appraisal, exactly the way she does after a ride up Madison to the hair salon, where she sits quietly. We both meditate upon her face and José's choreography of the dryer, brush, and Phoebe's long dark-blond hair.

"Now I don't want to go to Connecticut," I tell Lueza the next morning, "there's going to be one of those confrontations. They have them on the *Jerry Springer Show*. Like the public circus in Rome. I don't know if I want to ride up to Connecticut just to be shouted at."

The worst capper of all the things I did to my children was leaving them in the bloom of their lives, in the pride of their time; when, above all, you want to show your parents who you have become and to share with them the glory of your own children—even, perhaps, to show them "this, you see, is how I would have raised me."

I am part of the first generation who expect to see our children into the beginning of what once would have been called their old age. We are showing them how to live on into realms of life where we'd once have been content to hang around watching our grandchildren once a week. I say this as if this is not one of life's most exquisite goals.

Jeremy picks me up at the Gelbs with Phoebe and with Blair, whom he is going to marry. I want to show them to Bruce and Lueza. The more I connect all the wings, all the branches of my life into one network, the less likely it is to go totally blank. "You'll see Jeremy and Phoebe and meet Blair," I tell Lueza. "Then if I say I haven't seen them yet this trip, some-one will say, 'But you did, Jill, I was there.'"

Jeremy is driving Johanna's Jeep. I want to buy things for the kids.

"You already gave me my TV," Justin says.

"That's progress," Johanna says. "Mom took an ax to our TV, wouldn't let us have one."

I thought of it as political action.

We all go to a market the size of England. Who is the painter who did giant food? Here are giant hamburgers, iceberg lettuce big as globes, giant ivory bread loaves, slices soft as cold cream, big cordovan apples, carrots long and orange, string beans, bricks of cheese, barges of scarlet meat and Caucasian fleshy chicken, tubs of butter, tanks of ice cream.

Johanna's son Justin is thirteen. I was thirteen when I hated my cousin Julius and being indoors. I wanted to have a ranch and to be the first Congresswoman in frontier pants.

Justin's eyebrows are thick and calm. He is silent on the ride home.

"So, how do you write your stories for school?"

"I write on a computer," he says. He's watching the highway. "You know, Mom, the only billboards are for cars, cigarettes, and alcohol."

"Maybe," I say, "we'll go to New York together one day."

"I don't like New York. It's like a big mall to me," he says, "and I hate museums."

"You just want to ride and think about stuff?"

"I guess."

As we drive along, Justin watches out the window and broods, inventing attitudes and approaches, considering pitches and catches, and keeping himself steady through the distracting chaos of his younger brother and sister. He watches out the window and broods, but doesn't miss what's happening in the family around him—like a pitcher who knows when the guy on second is planning to steal third. Justin knows when the kids are using him, using some of his energy.

Ethan, who isn't shy, invents distractions, shouting, goading, until Justin turns on him. Then he's quiet, tugs at my sweater and says, "Do you know they have a class at my school and they call it art and we can just draw?"

Alice, like Ethan, is crazy about new words. "What's sex?" she says.

I say, "It's a game adults play."

"No," Johanna says firmly, "it's something two adults do as a way of showing they love each other."

Riding with Alice's head against my arm, I see pieces of Connecticut life, of Anatole and Sandy, Bliss and Todd, of Lynn and Dick with Priscilla and Claire, of Marcia and Dolph, Holly and Roger.

When we come home, we walk together—the colors of Connecticut are shifting, fading. We slip down piles of leaves, toss them into the air, the children's faces shift as quickly as the leaves, which crackle like old hands reaching as we tumble through the woods up along the hillside. My memory slips between now and then, cuts between the cold air, the ash smell, the snap of twigs. Shots from old scrapbooks.

I expect the confrontation will come now. Ideal, while the kids are getting their costumes together. We've eaten. Johanna's sitting here in one chair. I'm at one end of the couch. Jeremy turns on the TV, lies on the couch, puts his head in my lap and goes to sleep. This, of course, is the point of the confrontation.

It's Halloween Night. The moonlight shines down as we wander across lawns and driveways, all open (no security gates), with the doorways marked by lighted pumpkins and banners. Bags are filled with Hersheys and Milky Ways.

Jeremy tells me before he leaves Phoebe with us for the night, "I can imagine how it felt when I said I didn't need you. Today I told Phoebe I'd always take care of her and she said, 'I'll be a big girl and take care of myself.'"

Phoebe is already ahead of me.

Now the kids are watching *Men in Black* in Johanna's living room— Mr. Smith and Mr. Jones protecting the Earth from the scum of the Universe. Johanna and I are in the kitchen. I take cups off hooks to wash them. "This one's white with chubby crossed legs."

"Gloria Cole gave me that," Johanna says, "careful, it's put together with glue."

Johanna has drawers of linens and on every surface, baskets, books, tiny china houses, and like her mother, masses of artificial flowers in the dining room, on the round table like the ones my mother's models held in their arms while they posed.

It's night. I am with all my grandchildren in Justin's room with the sapphire plastic bubble chair—as space age as Vada's blue diamond. Darth Vadar made from a kit glares at me from a small chest. Yankee posters on the walls—*Star Trek: The Next Generation*, covers of *Baseball Weekly*, Bernie Williams, Derek Jeter, Andy Pettitt.

I sleep on a Marimekko pillow and Johanna carries a Marimekko bag I found in London just like the ones I'd buy from Flax—or, no, Design Research, in the very late fifties, when Mondrian was the look you needed.

We watch Justin's TV games. The one I can understand is where we invent characters—vests and wristlets, gauntlets, helmets and belts, flicking through colors and sizes, heights and faces, flipping through names and races, character traits, like accessible, easygoing, fast, angry, slow and secretive, highly charged.

Then I teach them charades the way my brother and sister and I played it. The Director invents a scene and tells the Actor how to play it. Then the person playing the Producer guesses what it is. There is no Writer.

"You're not like a regular grandma," Justin says.

Most memories I'd apply to rapture were about falling in love.

I thought, too, that hard drive was full. I'd done all the roads to rapture. Memory charges up, like a good old battery, and here's an entirely new road to go. Showing them how to draw coyotes and telling them stories of how it was, and learning from them how to send cartoon roses on e-mail and listening to their stories of how it is.

Now Ethan and Alice are wrapped around each other on the floor in thick duvets, and I'm in the lower bunk with Phoebe wrapped around me and vice versa, Justin thrashing in the bed above us. The breathing around me settles, hushes into sleep. They are my memory's link to eternity.

*T*hink about this: exercising memory as heavy aerobics.

A complex mnemonic goal was to envision all the planets, all the constellations, including astrological passages, all surrounded by wheels holding an encyclopedic array of arts and science knowledge. All of these concentric circles, an atrium of memory rooms you are looking down upon, observing the details of how the planets, the tides, the seasons, all the forces impact upon each other.

—Adapted from *The Art of Memory*, Frances A. Yates, 1966

I have just about finished my memory book. Then Stuart suggests I go back to talk with Zilkha, the doctor who said I would write again. "He'd be interested in hearing how you've done."

"You could tell him," I say. My attitude about doctors is a little like shopping. If you don't go into the store, you don't buy something.

We are in Zilkha's office. As I said, you say hello to a doctor and right away he wants to see you. "There's a new twenty-four-hour telemetry test, which will tell you something interesting," he says.

"I'm finishing my book," I say, "I really don't have time."

I hate being in hospitals. Hate tests. Tests also always lead to something else.

"I have an interesting book you should read, *Women and Epilepsy*," Zilkha says. "Let me see if we can set up the test next Tuesday." Zilkha asks his assistant to find the book. "You can look it over while I make the arrangements."

The assistant hands me the book. Everyone who works with Zilkha is crazy about him. It's like being a runner for Cicero.

"I can't loan you the book," Zilkha says, "but if you see anything you'd like to think about, we can copy those pages for you."

Stuart and I sit in the waiting room. He picks up *Vogue*. I flip through *Women and Epilepsy*.

"Is it interesting?" Stuart is watching me.

"Too much," I say. I note the pages I want to read again.

"It's the temporal lobe I want to see," Zilkha says when we've gathered again in his office.

I fear I know what he means. It's suddenly all too simple, all too elementary, my dear Zilkha.

We are driving to the Cromwell Hospital, through Hyde Park past children gathered for pony lessons. The black taxi is driven by a young woman. That's how long I've been in London. Black taxis remind me of our old black lunchboxes. My mother told the governesses they were classic. The bright colors on the lunchboxes Nona and Binnie had were painted with enamel: (a) to which we were allergic, (b) were made by a right-wing company, or (c) by a company that didn't hire Jews or blacks.

The park is sunny and I can see how appealing it would be to walk here today. That's also how long I've been in London.

Dr. Zilkha has the wizard's gift of sudden appearance. I've just hung up my jacket in my hospital room, put this blanket Jeremy sent me on the bed, and Zilkha is here.

"Did you go over the *Women and Epilepsy* pages?" he asks.

"Not yet." I have them with me, "but when I feel this taken care of, I don't think the brain's going to leap up for attention."

"But I do expect we'll see some abnormalities," he says.

I'm here to have an MRI test, followed by the new telemetry test, where, as I write, draw, or sleep, my brain will be on camera for twenty-four hours. Will this brain offer up another clue?

Stuart has left. He has written "Stuart loves Jill" on the white note-board on the wall opposite the bed.

The MRI test is simple. I lie down on a surgical table with my arms across my chest. A glass mask is clamped over my face. I'm slid into a giant cannon, so that Con Ed can test the sound impact of new drills for forty-five minutes.

Now I am back upstairs, alone in a beige room with beige built-ins and a pale, sea-blue cover on the bed. I see a jet crossing a rare sunny March sky above rows of large West Kensington Victorian houses. Around the corner Andrea Tana is painting. She is using exactly this shade of blue. If you painted on Wimpole Street, you wouldn't get this color as a reference point. This is why over in Wimpole Street territory one tends to draw in black and white, like Laurie.

Tracy is a neurological technician from Australia. She is gluing electrodes onto my head and asking me questions about what she calls "events." How many? Does it feel like sexual release? Any kind of release? How long do they last? Grand mal?

The electrodes are thirty-six-inch-long thin cords in sky blue, grass green, dark pink, lavender, yellow, and pearly gray. They are stuck to my scalp with glue dabbed onto their silver tops. I have become my own telecommunication system.

I could wear this, plug it in to e-mail, and you'll know what I'm thinking. Never mind the writing coming through my hand, sharing, showing you or letting you catch the sound of it with my voice. We will wave it across to each other. I am preparing for when you live on Mars and we'll exchange signals.

Are my brain waves unique?

The box by my side is sending signals to the screen in the nurses' watch room. The box is set into exactly the kind of white shoulder bag I'd like for a trip we're going on. It holds the key to this brain.

I am being watched as I study and write for twenty-four hours. There was no test like this ten years ago. This may not fix anything, but knowing how deep my space-outs go will help me know how safe I am, how safe my memory is, and most of all, how safe I am for a kid to be alone with.

I am to keep a diary of what happens. Three columns: the time, what I'm doing, and the Event.

I think of prizefights, charity dinners, and premiers as Events, but these have one similarity, a singularity in one's day. They are notable. Has my brain elected to have these events in order to grab attention fast? The heart-wrenching subject of the charity (far more appealing than the woman going to lunch); the powerful victim/hero of the prizefight, triumphant through pain; the panicked, fenced-in star.

Can the brain will trouble on itself, as it can pitch the body down the stairs? Can it make an Event without a committee?

My guess is, if Freud hadn't been left behind some time ago, that it all started when I saw my father leave his typewriter and come dashing to me when I was sick. Just as he dashed to his mother. Do heart attacks come from a warmer emotion than attacks of the brain?

After Stuart leaves, I decide I'm not going to use the phone tonight. I'm not running from what I've got here or from what I can learn, even though I can no longer trust convenience amnesia.

I read about a twenty-year-old woman who "from the age of four had had seizures due to a right temporal lobe astrocytoma. The attacks were essentially paroxysmal sexual manifestations, which could be triggered by fantasies in which the parental image appeared" (adapted from M. R. Trimble, ed. *Women and Epilepsy,* 1991, ch. 13).

So I wasn't evil when I'd have this indecent dizziness and imagine my parents making love, or late at night, after leaving my mother's room, I'd swoon, ostensibly turned on by the vision of her in her chiffon gown. These were seizures, which I thought were some kind of punishment for undressing everyone in my mind. It was the other way around. The mixed-up electric charges in my brain generated the erotic imagery and responses.

I am really looking. I feel anger, like when I realized I am an alcoholic. I am furious. "Why didn't anyone tell me?"

Images from the past tumble over. I hold them here. Freeze frame: the time I found myself thrown by the horse, certain I'd seen it buck, startled by an image of a naked man who suddenly appeared in front of me. I knew he wanted to rape me.

Sexual seizures associated with temporal lobe foci assume various forms. An eroticized aura may consist only of a diffuse sense of heightened arousal.

—*Women and Epilepsy*

I was quickly taken out of the school. Another time: I was driving down from Palo Alto on the Coast Highway. I had this wave of longing. Then I saw a naked man driving in the car next to me, handsome shoulders and neck. I woke up in a hospital. I was told I'd driven head on into a truck.

I could spend a lot of time listing incidents and reciting denials, which have nothing to do with my life today. "Denial," the book says, "is common and may be constructive, leading to planning for the future.

"It may also be destructive," says the next sentence. I thought of all the years I drove with my children in the car.

I look over Table 5.

Table 5. Adolescent epilepsy coping mechanism
Insight/acceptance—ideal, but unlikely
Denial—common
Constructive—allows development, planning
Destructive—increases non-compliance
Intellectualization—permits mastery
Rationalization—uncommon, fatalistic
Regression—common in early adolescent
Reaction formation—mastery
Projection—anger, frustration, alienation
Withdrawal—poor, promotes non-compliance
Panic—outbursts, diffuse hysteria
Acting out
Temper outbursts, aggression
Manipulation, non-compliance
Running away, verbal abuse
Flirting, suicide gestures
Drug abuse

—*Women and Epilepsy*

These mechanisms are not a lot different than addict and alcoholic coping mechanisms. But then, does anyone go through adolescence easily?

So my brain has a bad paste-up, like "what's wrong with this picture." I am not, then, essentially mythically evil.

I'm far more interested in the subject of memory. I flip the electrodes, hanging from my scalp like pastel dredlocks, over my shoulder, lie down, and read along in Frances Yates.

My mind eats up the stories, characters parading around the stage of this vibrant study of the time when memory's art was a discipline. Just as in my childhood, learning by rote was a discipline. We repeated multiplication tables until we remembered them—one of the vestiges of the art of memory. Today our children and grandchildren have calculators; their memories develop in new, audiovisual ways.

To excel at the art in earlier times, you'd devise systems filled with planets and zodiac signs, galaxies and mythic characters. Memory images could be comic—Furbies would fit ideally into memory images designed by a man named Publicious. One of his was a character called Unhappy Envy. (Is there a Happy Envy? Useful Envy?)

Guilio Camillo, famous in the sixteenth century, built a wooden memory theater with a working secret only the King of France, who paid for it, ever knew. Camillo dictated his theater's plan, with its seven concentric rows and seven aisles standing for the seven planets' attitudes, in seven days.

I far prefer imagining these complex plans using mythological figures as memory objects than looking at the stark, simple truths about my memory fault and its impact on my life.

In 1600, the Renaissance philosopher Giordano Bruno was burned at the stake because the church feared people would remember his buoyant, illustrated memory system.

Did I have to travel as far as I did? Could I have summoned up the images of my own past simply through the occult system of movies, pictures and old legends I have in books and notes around me? By flying miles back in time, I spun my mind back years. And now, is this wired-up twenty-four hours an allegorical event for my memory? No more nor less than reaching High Table at Oxford, when only a few months before that I couldn't make it home from the corner. Only five years ago, I couldn't have spent twenty-four hours alone with my memory. It would not have found enough to do.

The sun is down, the traffic's muffled and soft, like a dark cloth has settled round it.

From the Middle Ages there was "a change in attitude to the imagination," and classical memory systems were being regarded as "artificial memory." Memory was no longer part of Rhetoric. Man, the Renaissance said, could grasp a sense of a Highest Power behind his own image, beyond appearances. Maybe the Highest Power was the Scientific Age, when ideas no longer wore outfits; were no longer set around in galleries of niches, long and shadowed deep plum, like the windows on the dark sand Victorian brick houses I can see from the hospital window across from me. I could turn the windows into niches and put one of the stories I want to remember into each one. But I've got them here in this brain with me.

Just as the neurological department isn't certain exactly where memory rests in our brain, Frances Yates isn't sure where memory belongs as a study.

Is it a philosophy, psychology, or is it, in fact, an art? Or, as she seems to prefer, is memory a part of the soul and, therefore, is it part of theology?

Maybe memory is everywhere.

The sky is softening down to the color of pale gray Armani suede sneakers. A friend of mine says, "Why do you put brand names in your books? It dates them."

Exactly. Brands become mnemonic devices. Tell me Lanz and I see Josie, and I remember I forgot to mention the barbecue Jeremy had when I was in L.A. "Mom," he said, introducing me to a tall, blond young man, "this is Tim Davis." Tim is Josie's son. I'd know those eyes anywhere. Tim is a writer; Jeremy is his agent. Do we need to say a word about life's circles?

But I was thinking of suede. That's what the brain in the glass case in Professor Steven Rose's office looks like, butterscotch suede. I walk to the window. I can hear, by the sound of the cars, the last people leaving work. At once more determined and more diffuse, the traffic is background music to the point and tuck of my brain's moves across the screen, telling its long and curious family love story. But the road has always been a part of the story. The only thing I don't get is why there are miles of empty space.

Barry Gordon prefers the "contextual" organization of memory, that memories are linked by constellations formed around landmark events.

That's how it is when I hear a piece of an old movie score, and I am on the set watching Gene Kelly and Cyd Charisse whirl across the sound-stage. Then I'm at Emerson Junior High, watching Carol Rosson and Bob Redford dance in moonlight shining through the windows of the gym. What was it, "Mr. Sandman"? And I see Sandy Burns and Warner LeRoy dancing at Chadwick. I remember some of the words of the songs, and even more about what we were all doing and what our lives and our parents' lives were like at that time.

Would I ever fall in love? I remember my father, dancing stiff in his back brace, and my mother, her cigarette on her forefinger; how much love can go through and still look like that after all.

When I started writing again about old L.A., I'd play tapes of MGM movie musicals, and classical albums like the ones my mother played in her studio. Like a switch flicked, the time comes on. When I want to remember these last twenty years, I play Stuart's jazz albums and the music we danced to.

Long-running love is a rare thing, and as time goes by and you realize there's inevitably less time ahead than behind you, all the "memory" songs tear your heart out. But they always did; from the moment there's life, you've got loss: "Try to Remember," "Memory." I hear Barbra's voice and I can see her dancing under the menacing helicopters, at her wedding to Jim Brolin, in her garden when the lavendar roses reminded me of the wisteria shading my sister as she glided through the arbor on my father's arm at maybe the last event in our house. The images spin into a montage of the weddings of all my children.

I see the edge of light, like a last thought, above the rooftops, scalloped behind the chimney pots, fading up into a steadily dimming sky. Does my mind dim down with the light? I used to be a night writer. I want to keep awake with my brain tonight.

Remember my version of Steven Rose's memory test? This is how many I remember today (I looked at the test twelve hours ago!): lemon, orange, feather, leaf, cherries, safety pin, key, button, scissors, pushpin, paintbrush, paint, screw, nail, pencil (blue), scissors, shell, thread, acorn, pen, and book.

This is nineteen out of twenty. Fifteen is considered good. Fifteen is what I scored last night. But nineteen is what stayed with me. I remembered because I put the things in particular rooms in my mind. The paintbrush and paint, the pencil and the feather are in my mother's studio. The cherries, leaf, orange, lemon, acorn are in the kitchen. The safety pin, shell, button and thread are in my bathroom. And the pushpin, screw, nail and scissors are in the toolbox in Stuart's changing room. The book and pen are on my desk. I forgot the key. I always forget my key.

I have not remembered them in order exactly. I have to go up and down the stairs in my mind to find what I want, but at least I can remember most of it.

I have worked my art. Giving and receiving.

We now remember truths (and things that aren't so true, things that shouldn't be true) through our cinemythology. We remember war from war movies. We worry that some may make it funny, too readily heroic or too clean, because the sharp images we see repeated become the mythic images.

So should I say how quickly I remember people when I ask their signs? It's been a favorite mnemonic device again for the past thirty years, really, something like the magician's mnemonic games with cards, which suggest numbers. The signs suggest personalities we already know how to remember.

The sky's royal blue. Morning is here, and I exercise hard, jogging in place. My brain paces along steadily.

I am worried. What if a seizure doesn't come? What if they're watching and it doesn't happen? What's wonderful is I am not alone when my mind is with me. So maybe the MRI will show something in the temporal lobe. The issue over why I have lost some periods of memory and not others is not going to be so clear.

One theory is that memory is split into files by a psychological event, stored separately. This would explain why the events right before and after the swimming pool scene at Champney's don't register, but not why I've also lost so much about the eight to ten years before the swim.

But then, this theory also considers that the memories you acquire in one personality state are stored one place, and those picked up in another condition are stored elsewhere.

Of course. This is why I have almost surreal clarity of my memories of my childhood and of the sixties, when I was given the white powder called ephedrine, presumably for asthma, but, now I think, as much to counter the effects of the phenobarbital I was given every night. This, I read tonight, was the standard treatment for epilepsy when I was a child. The thing they didn't understand was addiction, and the catch that the very thing that stopped the seizures also triggered abuse.

The clarity changes, like a focus being adjusted, after a space-out, while the memories of the last years in Connecticut and in England are softer in the tones of monotone, an early russet and aqua color film.

The split personality idea is obvious and I think it's a condition any performer would understand. It just may be more extreme. When I'm quiet, writing and drawing, I go back to my best childhood times easily.

But when I'm in a group being funny, being "on," I lose all memory, go into routines. If I stop the "routine," the "riff," to really remember, to reflect on the people around me, to hear them so I can remember them, so I can write about them, I lose the riff. I can't go back into it. This would explain why, if you meet a performer "out of sync," she'll seem cold, distant. It's as if you put on outfits. I have trouble taking off the coat. Or is it a mask?

Kids see it. They're used to dressing up. That's why I'm far more at ease with them.

I can remember last night. I can remember what I have read.

The doctor and Tracy come to see me again. "How's your diary?" she asks.

"Nothing happening," I say. It is empty under 'Event.'

Zilkha says, "We haven't read the chart yet."

I have a feeling he already knows the answer.

Then at 10:05 in the morning, the Event comes, the gathering in, the looming fear. I look in on spinning men lying on their backs, legs open like Ziegfeld dancers, centrifugal force, a fugue, big color sheets blown out, scarlet, marigold and cobalt blue, clipper ships make a getaway, and I'm shuddering down, in, clattering down like the last tiger-eye marble rolling home across the glass.

Unlike some other people who smell and taste things, my space-outs are filled with color. Rats have no color sense, but they remember taste and smell. Birds remember color, but not smell or taste.

I turn to the book, to the page I've marked with a red tab. Here's the description, what Zilkha wanted me to see.

> A sudden libidinous feeling, sometimes associated with a sense of familiarity . . . as if she had suddenly been cut off from a sexual experience. There is no resolution.
>
> —*Women and Epilepsy*

So they have something to see, and it no longer terrifies me. We are sprung from our own land after all, and it is only fitting my brain should be built round a fault.

This twenty-four hours alone with my memory here has been an allegorical trip, as I've looked at all the ways I've trained it, pushed it, encouraged it and set it off, never sure it was there for the trip. It took me to the corner, to Oxford and to L.A., where what I really found, looking for the Polanski girl, was recognition of my own losses, which had little to do with sixties' melodramas and much more to do with the recognition of emotional responsibilities, of academic and political ambitions I tossed aside.

In these last years, for each image of physiological intrigue in Steven Rose, for every dense line of historic passion in *The Art of Memory,* I have faced another connection, which might well be a paragraph of *Women and Epilepsy.* And another sentence from an article by Peter Wolson, who talks about "Adaptive Grandiosity" (*Psychoanalytic Review,* 82[4]:).

Woolson, unlike a lot of analysts, says artists do need a powerful belief in their creative power to learn. It isn't easy for a young prodigy, for example, to focus when she, or he, is surrounded by maestros who know they'll never achieve what she will, or to function with that peculiar outsider's sensitivity in a world of curiosity, envy, a world waiting for the "hitch." If you don't have enough of this particular confidence that keeps you certain you're the best, no matter that you still have to practice or revise, you might be afraid to give yourself enough alone time for the talent to flourish, the ideas to evolve. Without this steely assurance, you might get into a room with a piano, a paintbrush, or a pen, and say, "Why bother?" Adaptive Grandiosity is the gut instinct that the world needs what you're doing.

I've always wanted to be unique. I hate when I don't get the table I want at the restaurant. (And it is never the first table.) At the same time, I want to be really liked; I ask a dozen times, "Do you think this works?" or "Should I wear this?"

Love is a man who will take you back so you can change your outfit. Stuart has Adaptive Grandiosity. He says, "This is what I'm wearing."

Most importantly, I've faced my own family, seen my parents and my children, and their children. What I lost, and what is there to find, to hold onto.

In the seventeenth century they worried that the Art of Memory was over. As, step by step, I've picked up my past, I wondered if there was anywhere to go. Could I remember more, and what would there be? What they did in the seventeenth century was apply aspects of *The Art of Memory* to these new explorations in the scientific method. So, here in this hospital, seeing the action in the temporal lobe of my brain, is the answer to these seizures, these gripping auras. But what I do with any answers I have found, how I apply them to my life, to time with my family, to working with my writers, is what matters.

I have applied this new memory muscle to the consideration of memory in all its guises and to aspects of my life I've never wanted to see. And I have learned that the only old idea about looking at the past is that there's nothing new to see.

The Englishman is sitting at the ash desk in his tan shirt. "What movie do you want to see tomorrow?" he asks.

I tell him.

"We saw that a month ago," he says.

Sometimes I'm not sure if that's true, or if he just doesn't want to see it. I do not want you to think I have a perfect memory for all things at all times.

"I told Judith we would have dinner with her Friday night," he says.

"Can't," I say, "we're going to Corinne's. It's Sabbath."

"I guess," he says, putting his hand on the arm I've wrapped around him, "I guess I forgot."